T0149569

HULCHUL:
The Common Ingredient of Motion and Time

SOHAN JAIN

authorHOUSE®

AuthorHouse™
1663 Liberty Drive
Bloomington, IN 47403
www.authorhouse.com
Phone: 1-800-839-8640

© *2012 by Sohan Jain. All rights reserved.*

No part of this book may be reproduced, stored in a retrieval system, or transmitted by any means without the written permission of the author.

Published by AuthorHouse   04/03/2012

ISBN: 978-1-4685-6298-9 (sc)
ISBN: 978-1-4685-6296-5 (hc)
ISBN: 978-1-4685-6297-2 (e)

Library of Congress Control Number: 2012904673

Any people depicted in stock imagery provided by Thinkstock are models, and such images are being used for illustrative purposes only.
Certain stock imagery © Thinkstock.

This book is printed on acid-free paper.

Because of the dynamic nature of the Internet, any web addresses or links contained in this book may have changed since publication and may no longer be valid. The views expressed in this work are solely those of the author and do not necessarily reflect the views of the publisher, and the publisher hereby disclaims any responsibility for them.

CONTENTS

PREFACE

I was enjoying my summer vacations after successfully completing the first academic year of the two-year Master's program in Mathematics at Delhi University. During the two months of summer vacations, I had plenty of time to enjoy my favorite pastime of reading books in the university library after a hectic year of being a full-time student and having a full-time job teaching mathematics to high school students. Getting a temporary job as a mathematics teacher for one academic year was a lucky break for me since I had just a bachelor's degree in mathematics (an additional degree in education is a normal requirement for school teachers in India). This occurred because of a temporary shortage of mathematics teachers in Delhi at the time. It was hard work but I received the full salary as a teacher! This gave me a temporary respite from my constant financial struggle. I was in a good mood.

For the second year, I chose most of the courses in applied mathematics, including the Theory of Relativity and the Theory of Elasticity among other courses. I learned physics mostly through mathematics. I was excited about the Theory of Relativity in particular. The Theory of Relativity, Quantum Mechanics and Einstein were some of the buzzwords at the time in the academic world. I wanted to choose Quantum Mechanics, too, but after browsing a couple of books on the subject, I dropped the idea as I found it involved too much physics. I was anxious about the Theory of Relativity, too. I browsed a few books on the subject and tried to get a feel of it but I could not. Why this theory was called the Theory of Relativity was puzzling to me. At that time, it appeared to me more a philosophy of certain newly discovered physical realities.

To be ready for the first day of the class on Relativity, I waited for the availability of textbooks in the textbook section of the library, which were made available a week before the classes would begin. Text books were only available for use in the library; one could take them home only overnight.

I borrowed the textbook on the Theory of Relativity on the very first day it was made available. The book seemed to be well written and to the point. I was never spell bound by a non-fiction as much as I was by this book. (Unfortunately, I remember neither the author's name nor the title of the book now.) It had the usual material starting with Michelson-Morley experiments and different reactions and interpretations of the surprising outcome of the experiments by the contemporary scientists including Einstein.

I was fascinated most by the concept of **time dilation**. "What an abstract and elegant design of the working of the universe! Time dilation has to be there!" I said to myself. I felt liberated by the idea that I could travel to any part of the universe in my lifetime and I was no longer stuck on the planet Earth! Ironically, at the time, I had never traveled more than 40 miles away from Delhi and I was already 21 years old!

I read the book cover to cover in the next two to three days, reading some parts of it in details and browsing the other parts. However, I was turned off by some other conclusions like the twin paradox, time dilation depending on relative motion only, direction and certain other premises of the theory. I did not know the reasons why I believed so at that time. I think I know some of the reasons now, more than four decades later: Time dilation should not depend on relative motion alone. Even without moving on a spaceship, we should be in a state of time dilation due to motion of the Earth, the Solar System and the Milky Way. In fact, all bodies should be in a state of time dilation with varying degrees. I felt very strongly that there was **"more to time dilation"** than it appeared.

In the last four decades, not a single day passed when I did not think, at least for a few minutes, sometimes for hours, about time dilation in particular and time in general, and an intrinsic and deeper relationship between time and motion in the light of time dilation. However, I was in no hurry to complete my research work, it was my best pastime. Naturally, my first priority was my family and my full-time job in computer software development, which provided my bread and butter.

Only in the last few years, I realized that I must complete my research work as no one can live on forever. I retired from my job. As a result, I found plenty of time—quality time—for my independent research. I started documenting the research outcome formally on my laptop. Before this, I just used to jot down my ideas and thoughts in composition note books—I have already filled 12 of them. I am presenting this formal documentation of my research work in the form of a book as another perspective of motion and time in the light of time dilation with a necessary mathematical framework. I do not know whether the physicists and mathematicians will find the contents of the book of any value or shun it as naïve. However, I enjoyed every moment of my research when it was frustrating and when it was productive.

Sohan Jain

March 2012, USA

thehulchulbook@gmail.com

INTRODUCTION

This book is the outcome of an independent research work done by the author at a slow pace over a long period. This started with a quest for an intrinsic, deeper relationship between motion and time in the light of **time dilation**. The premises for the quest were:

- **Motion is not continuous; it is discrete**. If a body moves for an hour, does it necessarily move during each instant of time during the one hour? We do not believe so. Motion of a body is divided into indivisible, smallest possible movements of the body; we will call these movements **prime ticks (p-ticks)** of the body. Prime ticks are to motion what elementary particles are to matter, and what photons are to light.

  As a consequence of the above premise, some other physical phenomena may not occur continuously; they occur intermittently with varying fineness of intermittency. For example: An electron shared by two atoms may not be shared continuously; it may be shared intermittently by the two atoms.

- **All kinds of motions are the same in terms of prime ticks:** This includes: translational, rotational, vibrating, oscillating, multi-directional (explosive, implosive, waves) motions or any combination of these. Example: Internal and external motion of clouds or bio-psychological systems.

- **A possible relationship between motion and time:** Motion of a body is made of prime ticks. Prime ticks are events. Events have an order and concurrency. The concurrency of prime ticks, as events, implies time. This implies a different kind of relationship between motion and time, possibly the kind of relationship we were seeking.

- **There is time only because there is motion**: The above relationship between motion and time implies that there is time only because there is motion.

- **Instants of motion, instants of time and time outage**: Just as time has instants of time, motion has instants of motion, too. An instant of motion of a body is possibly the movement of the body during an instant of time; this appears to be the same as a prime tick. Instants of time and motion can be divided into three classes: pure instants of time, pure instants of motion, and composite instants of time and motion. If an instant of time and an instant of motion are concurrent, then it is a composite instant. Otherwise, they are pure instants of their respective classes. The sequences of the three types of instants are interspersed into a single sequence of their occurrences. A body does not experience time during pure instants of motion, a phenomenon we will call **time outage.** Time outage is not continuous; it is intermittent. Time outage appears to be the cause of **time dilation**.

** Figure (Sequence of instants of motion and time of a body)

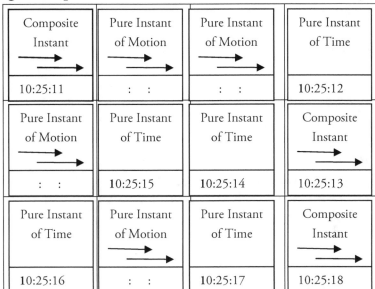

- **Inner-outer body relationship:** If body b is a constituent part of another body B, then b is an inner body of B, and B is an outer body of b. We will denote this relationship by: b ⊂ B.

- **Internal and external motion of a body:** Suppose b ⊂ B. Internal motion of a body B is the motion of all constituent parts (inner bodies) of B relative to each other. If b moves in B, then it is an external motion of b and a part of an internal motion of B. Internal motion and external motion are two views of the same motion.

- **Inheritance of motion:** A body b goes wherever its outer bodies go. An inner body inherits the external motion of all of its outer bodies. An outer body inherits the motion of all of its inner bodies.

- **Internal motion of a body increases outer-body-ward and external motion increases inner-body-ward:** This is due to inheritance of motion. This has a striking similarity with time dilation—time slows down as the speed of motion increases. In terms of p-ticks, the speed of motion is the same as the *intensity of p-ticks*. External motion of a body b is actually the motion of b and internal motion of b is the time for the body b as we will see in Chapter 3.

We will build a mathematical framework to accommodate the above concepts and more to put motion and time in a different perspective in the light of time dilation. This will include a theory of concurrency of events with focus on prime ticks as events. Some of the hulchul related concepts possibly explain the whys for:

- The principle behind the working of a particle accelerator.
- Compartmentalized structure of atoms.
- Fast moving elementary particles have a wavy straight motion and not straight motion.
- Principle of Relativity (motion part of it).
- Speed of light is independent of the source of the light.

Structure of the Contents of the Book

- **Chapters 1-4**: The four main chapters.
- **Chapter 5**: This contains summary of all hulchul related concepts and conclusions.
- **Sections/Sub-sections:** Each chapter is divided into sections, and a section may be divided into sub-sections with unique IDs.
- **Figures, Tables and Equations/Inequalities:** Each figure has the ID of the subsection the figure belongs to; this ID may be suffixed by one letter to make it unique. The same rule applies to tables and equations/inequalities.
- **Questions:** Throughout the book we ask questions that do not appear to have easy answers—questions that arose in the mind of the author or are likely to arise in the minds of readers.
- **Notes:** Throughout the book, we have short notes where situations warrant extra explanations.

At the end, we have:

- **A list of mathematical notations related to hulchul.**
- **A list of figures.**

- **A list of terms related to hulchul**: A list of all terms introduced for the sake of hulchul in their conceptual order.
- **A glossary of terms**: Definitions of all terms introduced for the sake of hulchul or interpretation of some other concepts used in terms of hulchul in alphabetical order.
- **Bibliography**
- **Indexes**.

CHAPTER 1

Prime Ticks: The Building Blocks of Motion and Time

1.1 What Is Hulchul?

We believe that motion and time are two sides of the same coin; the coin, here, is **hulchul**. What is hulchul? **Literarily**, hulchul is a common Hindi word, an abstract noun that means movement, commotion, hustle-bustle, agitation, any activity attracting one's attention, with a wide variety of its connotations. Of its two syllables, "hul" means shaking and "chul" means moving; both syllables rhyme with the word "null." Some examples of sentences using the word hulchul:

- Are you having a party? I see some hulchul in your house.
- He had hulchul in his heart after he saw the woman of his dreams again.
- There is a lot of political hulchul before a major election.
- There is some hulchul here mostly on Friday evenings; otherwise, it is a quiet shopping center.
- We had a couple of earthquakes in the last year; Mother Nature is having some hulchul, too.

All **hulchuls** in the above sentences imply some kind of motion directly or indirectly—all are complex motions. This does not mean hulchul implies only complex motion—it includes any kind of motion, simple or complex.

1.2 Instants of Time and Instants of Motion

Goal: Our goal in this book is to explore for an intrinsic, deeper relation between motion and time in the light of time dilation.

Further, to establish a mathematical theory of motion and time to demonstrate: Just as time is a sequence of instants of time, similarly, motion is a sequence of instants of motion; instants of motion are the smallest movements, indivisible any further. Hulchul is **just** a set of these small movements we will call **prime ticks**. The two sequences of instants of motion and time are interspersed into a common sequence; that is, the common sequence is a mix of instants of motion and instants of time. Alternatively, the common sequence is a mix of three types of instants: Pure instants of motion, pure instants of time, and composite instants; a composite instant is both an instant of motion and an instant of time. We believe:

*The instants, making the common sequence, are the common ingredient of motion and time; further, a body does not experience any time during pure instants of motion—a phenomenon we will call **time outage**; time dilation is the same as a time outage; **the way a time outage is arrived at, explains why there is time dilation**.*

1.3 Bodies, Motion, and Time—A Different Perspective

1.3.1 Motion and Time

Motion: For the purpose of this research, hulchul refers to any kind of motion that may be translational, rotational, vibrating, oscillating, etc. or a mix of them. Here, motion and time are two views of the same physical concept we will call hulchul; we will define hulchul, with a mathematical precision later. For the moment, to keep it simple, we will assume motion of a body as a sequence of small movements, indivisible any further, of the body.

Motion is not continuous. Motion is like a movie—a sequence of individual picture frames; individual picture frames, here, are the instants of motion.

Time: One common use of time is to measure the speed of a moving body. The Theory of Relativity elevated this relationship between motion and time to another level called **Time Dilation**: The time for a body slows downs as the body moves; the slowdown of time increases as the speed of the body increases. One of the objectives of this research is to go a step further and to introduce the notion of **Time Outage** as alluded to earlier.

1.3.2 Inner and Outer Bodies, and Their Relationship

If b1 is a constituent part of a body b, then we will call b1 an **inner body** of b, and b an **outer body** of b1. The inner-outer body relationship is transitive. We denote this relationship between b1 and b as: b1 ⊂ b. The inner-outer body relationship is one of the fundamentals of the notion of hulchul.

The inner-outer body relationship between two bodies is not necessarily permanent or continuous: A spaceship on the surface of the Earth, ready for launch to orbit Mars, is an inner body of the Earth but after the spaceship begins orbiting Mars, it is an inner body of Mars. The inner-outer body relationship between two bodies may be **continuous** or **intermittent**. The preceding observation adds complexity to hulchul; hulchul deals with the inner-outer body relationship extensively. For example: Earth is an inner body of the solar system; compared to this, a double star and its planets have a more complex inner-outer body relationship. The inner-outer body relationship between an electron and two atoms sharing the electron is more complex than an atom having an electron permanently. Hulchul theory, by conception and design, is intended to handle this kind of complexity.

Lineal Bodies and Non-Lineal Bodies: Two bodies will be called **lineal bodies** if exactly one of the two bodies is an inner body of the other body. If two bodies are not lineal, then they will be called **non-lineal bodies**.

Two or more lineal bodies can be viewed as one body embedded into another body serially (not in parallel). The notion of lineal and non-lineal bodies is central to hulchul. Later, we will extend this notion to the small, indivisible *movements* of bodies, too. For example, in Figure 1.3.2, movements t1 and t3 are lineal; movements t1 and t2 are non-lineal.

** **Figure 1.3.2**

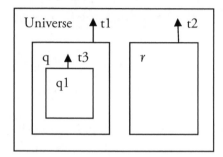

Bodies q and q1 are lineal.
Bodies q and r are non-lineal.
Bodies q1 and r are non-lineal.

Naming Convention for Inner-Outer Bodies: As much as possible, we will use the following names in the following order for bodies in a chain of inner-outer bodies:

$$\ldots b2 \subset b1 \subset b0 \subset b \subset B \subset B0 \subset B1 \subset B2 \ldots$$

1.3.3 Inheritance of Motion

We will assume the following characteristic of a body and its motion: **A body goes wherever any of its outer bodies goes; in addition to this, the inner body may move independently, too.** We will define this characteristic in this way: **An inner body inherits the motion of its outer body.** For example: Passengers on a moving train go wherever the train goes. The passengers and train go wherever the Earth goes.

Passengers, train, and the Earth go wherever the solar system goes, and so on. In each case, additionally, an inner body may also move independently of its outer bodies.

1.3.4 About Bodies—Their Four Types

For the purpose of this research, all bodies, in general, are the same and we will treat them generally the same way. However, we will divide the bodies into the following four categories:

The U Body—We will call the universe as a whole **the U body**. To keep it simple, we will assume that all bodies are inner bodies of U and U has no outer body.

Elementary Bodies—We will define an **elementary body** as one having no inner body. Elementary particles are elementary bodies (on the assumption that an elementary particle has no inner body).

Phantom Bodies—We will define a **phantom body** as a body that does not inherit the motion of its outer bodies; another characteristic of a phantom body is that it does not have any outer bodies except the U body. We will assume photons and light are examples of phantom bodies.

We believe it is the first characteristic of a phantom body as to why the speed of the source of light does not add to the speed of light. In fact, the source of light is not even an outer body of light. The two characteristics of phantom bodies are complementary of each other.

Strictly speaking, the above implied behavior of light and photons is correct when they are moving in a vacuum; when they interact with matter, they are in a more complex state.

We will discuss this more under Proposition #3.

Common Bodies—Any body, other than the U body, an elementary body or a phantom body, will be called a **common body**. Examples: atom, car, animal, heart, brain, planet, star, galaxy.

1.3.5 Outer Body as a Frame of Reference

For the purpose of this research, for the motion of a body b, we will use mostly an outer body B of b as a **frame of reference**. The outer body B can be turned into a practical frame of reference, for example, by firmly attaching a coordinate system to B so the frame of reference goes wherever the outer body B goes and it does not move in B. The body b goes wherever the frame of reference and the body B go; aside from this, the body b can have its own independent motion with reference to B.

1.3.6 Internal, External and Composite Motion

Internal and External Motion: We will divide motion of a body b into two broad categories: external motion and internal motion. The external motion of b is with reference to an outer body B of b whereas the internal motion of b is the collective motion of all inner bodies of b with reference to b. *Internal motion and external motion are two views of the same motion.* If a body b moves in a body B, then the underlying motion is the external motion of b and a part of the internal motion of B. The internal motion of a body is as critical as the external motion is to the notion of hulchul.

Composite Motion of a Body: Internal motion and external motion of a body are not necessarily mutually exclusive. A body b can have internal motion and external motion concurrently. We will call this **composite motion** of the body b. For example: a ship is moving and, at the same time, a number of other activities, involving motion, are going on inside the ship; some of the motion caused by these activities may be intermittently concurrent with the motion of the ship as a whole. Therefore, the ship may have: 1) both internal and external motion concurrently during some instants of time, 2) only internal motion during some other instants of time and 3) only external motion during the remaining instants of time.

1.3.7 Extended and Absolute External Motion of a Body

Suppose $b \subset B \subset B1 \subset B2$. Suppose b moves in B, and B moves in B1. Then b moves in B1 and this will be called an **extended external motion** of b in B1. If B1 also moves in B2, then b has an extended external motion in B2. Extended external motion of b in U will be called an absolute motion of b. We will create a formal mathematical framework for extended and absolute motion of a body in Chapter 3. (Internal motion of a body is always absolute.)

** **Figure 1.3.7**

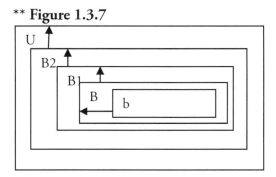

1.3.8 An Observation to Understand Motion and the Absolute Nature of Motion

Look at a book lying on a table. The book appears to be at rest but it is not. Internally, trillions of atoms and their constituent parts are frantically jostling with each other—we will call this the internal motion of the book. On the other hand, if we move the book on the table, we will call this an external motion of the book relative to the table. Instead, if we move the table, the book again has external motion though in a different frame of reference—the room. If we move neither the book nor the table, the book still has external motion for so many other reasons such as: the Earth is rotating about its axis and it is also moving around the Sun; the Sun is rotating about its axis and it is also moving in our galaxy; the galaxy itself is spinning; and so on until we come to the U body.

Therefore, the book has absolute motion in the U body aside from several relative motions. It has two kinds of motions: internal motion and external motion. In fact, the same can be said about any body such as: an electron, photon, atom, car, animal, heart, brain, planet, star, galaxy or the U body, though an electron and photon can have only external motion and the U body can have only internal motion.

Absolute Motion: We can talk about the motion of the Earth in the solar system, and motion of the Earth in the Milky Way. Can we talk, in a similar way, about the motion of Earth in the U body, at least, in the sense of a mathematical abstraction? Absolute motion of a body b can be thought of as the total motion of the body b relative to all outer bodies of b, which is the same as the motion of b in the U body.

Question: In view of the inheritance of internal motion by outer bodies and external motion by inner bodies, it appears that an outer body has more internal motion than any of its inner bodies has, and an inner body has more external motion than any of its outer bodies has. That is, internal motion increases **outer-body-ward** and external motion increases **inner-body-ward. In that case, is the sum of internal motion and external motion the same for all bodies?**

In fact, we will show that this is true if we measure the motion in a certain way—to be called **concurrency;** we will establish this in Chapter 3. We will also show that concurrency of external motion behaves like motion and concurrency of internal motion behaves like time.

1.3.9 Interchangeability of Internal and External Motion

A thought experiment: Let us do a thought experiment. Suppose we have a system of four point particles P1, P2, P3 and P4. We will call this a point system PS4. In Figure 1.3.9A, some or all four point particles move slightly in different directions, some of them may not even move at all. This causes an internal motion in the point-system PS4. Now suppose the four points again move slightly, and this time, all points move the same distance and in the same direction as shown in Figure 1.3.9A. This is a case of an external motion of the point system PS4 as a whole. As shown in Figure 1.3.9A, the four points move the same distance and in the same direction again and, finally, four points move in different directions.

Now suppose the four points undergo a little collective motion, say, 100 times; sometimes they may undergo an internal motion and sometimes an external motion. Therefore, we see that a body may sometimes experience internal motion and sometimes external motion. What we want to emphasize here is that internal motion of a body may change to external motion and vice-versa, from instant to instant, depending on whether its constituent parts, enough constituent parts making up the entire body, move the same distance and in the same direction or not.

We notice, so far, that if there is an internal motion, then there is no external motion and vice versa; but this is not necessarily to be so. Internal and external motion of a body may occur concurrently. We will call this a composite motion. A composite motion is visualized in Figure 1.3.9B where one of the point particles, P1, is replaced by another point system PS5, which consists of five point particles. PS5 is an inner body of PS4. PS5 behaves like PS4 in its own right. PS5 can experience composite motion when PS5 moves as a whole in PS4 and at the same time, PS5 has internal motion.

** Figure 1.3.9A

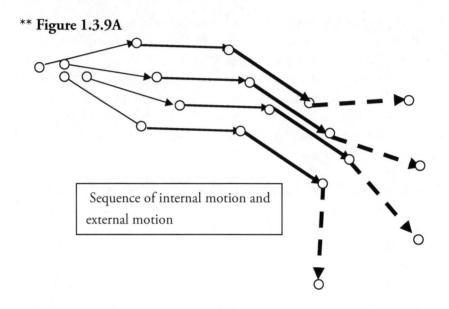

Sequence of internal motion and external motion

** Figure 1.3.9B

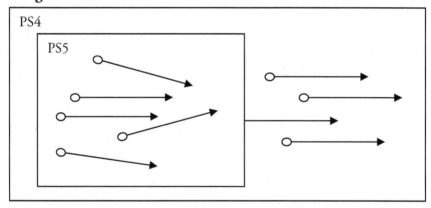

Question: On the assumption that electrons do move in an atom: Do the electrons in a shell/sub-shell of an atom move independently in the shell/sub-shell so a shell/sub-shell has internal motion or do they all move in sync, that is as one body, so a shell/sub-shell moves or wobbles as a whole around its nucleus? Or, is it a mix of the two kinds of motions?

1.4 Prime Ticks—The Building Blocks of Time and Motion

1.4.1 Ticks, Prime Ticks and Chained Ticks

We believe that any kind of motion of a body can be divided into tiny indivisible movements of the body; these movements are very similar for any kind of motion of any body. We will assume that motion of a body is not continuous; it is discrete.

Suppose b1 and b are two bodies such that b1 ⊂ b. We may perceive the motion of b1 in b, for example, as a sequence of the repetition of the following two steps:

- Step 1: The body b1 moves in b with b as a frame of reference.
- Step 2: After the movement in Step 1, there is a change in the motion of b1 in b. The change may be: b1 has a pause, and/or its direction changes.

We will call an occurrence of the above two steps a **tick of b1 in b**; we will say: **b1 ticks in b**. Let us name this tick as t1 and denote it as **t1 = t[b1, b]**. We will call the body b1 the **object body of t1** and body b the **reference body of t1.**

It is possible that there exists a body b0 such that b1 ⊂ b0 ⊂ b, b1 ticks in b0 and b0 ticks in b concurrently; then we will call the tick of b1 in b a **chained tick.** We will represent this as: **t[b1, b] = t[b1, b0] + t[b0, b] and we say: tick t[b1, b] is decomposed into two ticks t[b1, b0] and t[b0, b]**. A tick may or may not be decomposed into two or more ticks. If a tick t1 cannot be decomposed in the preceding manner, then we will call t1 a **prime tick** or **p-tick** of b1 in b. A tick is either a prime tick or a chained tick; the term "tick" is just a common name for prime ticks and chained ticks. (Technically, we can also say, prime ticks and chained ticks are sub-classes of the class ticks.)

A prime tick of a body is the smallest possible movement of the body, indivisible any further. Prime ticks are to motion what elementary particles are to matter or what photons are to light. Prime ticks are all-pervasive in the universe.

Suppose t1 = t[b1, b]. The tick t1 is an event and it occurs at an instant of time. If t1 is a chained tick, then the constituent p-ticks of t1 are concurrent.

** **Figure 1.4.1**

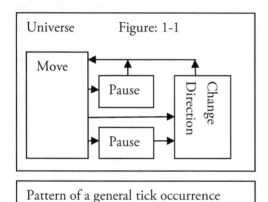

Pattern of a general tick occurrence

An Alternative Definition of Ticks

Suppose b1 ⊂ b. **The movement of a body b1 in b, during an instant of time, will be called a tick of b1 in b.** We will denote the tick as **t1 = t[b1, b]** where t1 is just a name given to the tick; we will say: **b1 ticks in b**. If there exists a body b0 such that b1 ⊂ b0 ⊂ b, b1 ticks in b0, and b0 ticks in b then t will be called a **chained tick**. If t is not a chained tick then it will be called a **prime tick**.

Difference between the original and the alternative definitions of a tick are:

1. The alternative definition explicitly links a tick to a single instant of time; the original definition does not though an instant of time is implied.
2. The original definition explicitly links a tick to a pause of the motion and/or a change in the direction of the motion; the alternative definition does not. A pause or change in direction is a transition from one instant of motion to another.

It appears that if a body moves during two consecutives instants of time, then the body changes its direction between the two instants of time unless the body pauses (momentarily or permanently). This may be reason why an elementary particle appears to have wavy straight motion rather than straight motion. The waviness of the motion depends on how fast the elementary particle is moving—faster the motion, more the number of p-ticks, more the waviness of the motion.

1.4.2 Empty and Non-empty Ticks

If b1 ⊂ b and b1 does not tick in b, then we will call t[b1, b] an **empty tick**. If b1 does tick in b, then we will call t[b1, b] a **non-empty** tick.

Empty ticks are a mathematical convenience and even a necessity. Therefore, there are three types of ticks: Empty ticks, prime ticks, and chained ticks.

1.4.3 Length of a Tick

We will call the number of constituent prime ticks of a tick the **length of the tick**. A tick is empty, prime tick or chained tick according as the length of the tick is 0, 1 or > 1.

Nth P-Tick in a Tick: P-ticks in a tick are ordered by their object bodies outer-body-ward. The very first p-tick of an external tick t of a body b will be called the **1ˢᵗ p-tick of t**; the nth p-tick of t will be called **nth p-tick** of the tick t.

1.4.4 External and Internal Ticks

If t1 = t[b1, b], then t1 will be called an **external tick** of b1 and an **internal tick** of b. External tick and internal tick are two views of the same tick.

1.4.5 Motion in Terms of Prime Ticks

Suppose two cars move on the same track, side by side, for an hour continuously—one car moves at the speed of 10 miles per hour and the other car moves at the speed of 100 miles per hour. In one hour, the slower car makes fewer prime ticks than the faster car does. The same rule applies even if the two cars move intermittently, that is, they do not move continuously. It does not matter even if the two cars are changing directions in any manner. Therefore, the motion of a body is slower if it makes a fewer number of prime ticks; the motion of a body is faster if it makes a greater number of prime ticks in a given time; it is similar to how the number of photons makes a difference between dim light and bright light.

Note: Earlier, instead of defining motion as a sequence of prime ticks where a prime tick is an occurrence of the two steps, we could also have simply assumed a **prime tick** as the smallest possible movement, indivisible any further. However, there is a point regarding why we used the "two steps" construct instead, as the following question suggests:

Question: Do moving photons and other elementary particles change direction rapidly?

If we assume that an elementary particle moves fast and therefore it may not pause frequently enough but it may change direction frequently, then that may be the reason why the path of photons or elementary particles is not **straight;** rather**,** it is **wavy straight**.

In this chapter, we will analyze only concurrent ticks, that is, the ticks that occur at the same instant of time. In the next chapter, we will analyze sets and sequences of ticks not necessarily concurrent. Now we will introduce different types of ticks and analyze them.

1.4.6 Visualization of Different Scenarios of Ticks

We will use some drawings to visualize different types of ticks using the following conventions:

- **Boxes:** We will use a box to represent a body. We will use two boxes one inside another to represent inner-outer body relationship between two bodies; this also represents two lineal bodies.
- **Bold line:** If a bold line connects an inner body b1 to an outer body b, then **b1 does not tick in b.**
- **Arrow with a bold line:** If an arrow with a bold line connects an inner body b1 to outer body b, then **b1 ticks in b.**
- **Arrow with a dashed line:** If an arrow with a dashed line connects an inner body b1 to outer body b, then **t[b1, b] is an inherited tick**.

Note: Usually, we will show arrows all pointing in the same direction (upward), though a tick can be in any direction in the three dimensional space.

At the end of this chapter, we will show a general diagram demonstrating all types of ticks.

1.5 Types of Ticks and Their Properties

1.5.01 Lineal and Non-lineal Ticks

Lineal Ticks: We will call two ticks **lineal ticks** if the reference body of one tick is the same as the object body of the other tick or the reference body of one tick is lineal to the object body of the other tick.

Two lineal ticks are always concurrent. Any two prime ticks in a chained tick are lineal. Later, we will show that two independent concurrent external prime ticks of a body are always lineal.

Example: Two or more lineal ticks can also be viewed as one tick embedded in another tick serially (not in parallel). If a passenger walks in a moving train, then this part of the motion of the passenger is embedded into the motion of the train; the passenger experiences both of the motions—sometimes, one or the other motion, sometimes both motions concurrently.

Non-lineal Ticks: We will call two ticks **non-lineal ticks** if the object bodies of the two ticks are non-lineal. Two non-lineal ticks may or may not be concurrent.

1.5.02 Composite Tick and Pure Ticks

Composite Tick: If b has an internal tick and an external tick concurrently, then we will say b has a **composite tick** and denote it by **t[(b)]**.

Pure Internal Tick: If a body b has an internal tick but no external tick concurrently, then we will say b has a **pure internal tick** and denote it by **t((b))**.

Pure External Tick: If a body b has an external tick but no internal tick concurrently, then we will say b has a **pure external tick** and denote it **t[[b]]**.

1.5.03 Property #1 of Ticks: The Decomposition of a Chained Tick into all Prime Ticks

A chained tick t[bn, b] can be decomposed into all prime ticks. In general, a chained tick can be decomposed into n prime ticks where n $\geq$ 2 and bn $\subset$ bn-1 $\subset$ bn-2 . . . . $\subset$ b1 $\subset$ b:

t[bn, b] = t[bn, bn-1] + b[n-1, bn-2] + ... t[b1, b]

Proof: We assume a chained tick has a finite number of p-ticks. By definition, a chained tick can be decomposed into two non-empty ticks. If the two ticks are prime ticks, then we have proven the property otherwise we repeat the process iteratively on the ticks that are chained ticks.

1.5.04 Property #2 of Ticks: Axioms of Inheritance of Ticks

We will assume the following two axioms without a proof. They are based on the inheritance of motion.

Axiom of inheritance of internal ticks: A body b inherits all non-empty internal ticks of any inner body b1 of b as non-empty internal ticks.

Axiom of inheritance of external ticks: A body b inherits all non-empty external ticks of any outer body B of b as non-empty external ticks.

Source and Destination Ticks: If the tick t1 is inherited from the tick t2, then we will call t1 a **source tick** and t2 a **destination tick**.

How to Differentiate Source and Destination Ticks: In the case of inheritance of internal ticks, the tick of the inner body is the source tick. In the case of inheritance of external ticks, the tick of the outer body is the source tick.

Notes on Inheritance of Ticks

1. **Prime Tick versus Chained Tick**: A prime tick may not be inherited as a prime tick; the inherited tick may result in a chained tick in case the body inheriting the tick already has a tick of the same type. The rule is: length of the destination tick ≥ length of the source tick.

2. It appears that the two inheritance properties above can be derived from the inheritance properties of motion, however, we will rather assume the two axioms of inheritance of ticks.

3. In general, a composite tick of a body b may not be inherited as a composite tick by an inner or outer body of b. However, as we will see later, a composite tick may be inherited as a composite tick in some situations.

Question: Does a destination tick **always occur concurrently** with the source tick? For example, a rope lying curled on the ground: if we pull an end of the rope, not all parts of the rope begin moving immediately.

We assume that the answer is "yes" in case of rigid bodies but we are not sure in case of bodies in the form of fluids, gases, or waves. It is possible that in case of bodies of a certain type, an inherited tick of the original tick may lag behind. In any case, we believe this concept (of inheritance of ticks) is worth pursuing. We may need to define some kind of "level of rigidity of bodies" for this purpose.

Notation for inheritance of ticks

We will use the following notations:

"<=" to represent "inherits"
"<≠" to represent "does not inherit"
"=>" to represent "is inherited by"
"≠>" to represent "is not inherited by"

Examples
Suppose b1 ⊂ b ⊂ B, t1 = t[b1, B], t2 = [b, B], t3 = [b1, b]
Then
t1 <= t2 means: t1 inherits t2 (t1 is implied by t2)
t2 => t1 means: t2 is inherited by t1 (t2 implies t1)
t1 <= t3: t1 inherits t3 (t1 is implied by t3)
t3 => t1: t3 is inherited by t1 (t3 implies t1)
This means: t1 <= t2 and t2 => t1 are the same.

** **Figure 1.5.04C** ** **Figure 1.5.04B** ** **Figure 1.5.04A**

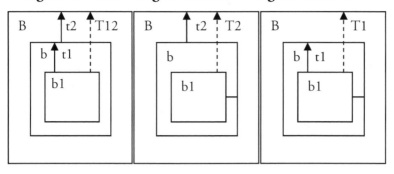

Figure 1.5.04A: t1 = t[b1, b], T1 = [b1, B], t1 => T1;

Figure 1.5.04B: t2 = t[b, B], T2 = t[b1, B,], t2 => T2

Figure 1.5.04C: T12 = t1 + t2 = t[b1, b] + t[b, B] = t[b1, B]
t1 ≠> T12.
t2 ≠> T12.
T12 is a chained tick.
t1 and t2 are prime ticks.

1.5.05 Range of Lineal Bodies of a Tick

Suppose b1 ⊂ b and b1 ticks in b. Let t1 = t[b1, b]. We will call the pair of object and reference bodies b1 and b associated with the tick t1 as the **range of lineal bodies of a tick and denote it by [b1, b]**. We note the following:

- t[b1, b] implies t[b2, b] if b2 ⊂ b1
  This is due to the property of inheritance of external ticks by an inner body.
- t[b1, b] implies t[b1, B] if b ⊂ B
  This is due to the property of inheritance of internal ticks by an outer body.
- t[b1, b] implies t[b2, B] if b2 ⊂ b1⊂ b ⊂ B
  This is due to the properties of inheritance of an external tick by an inner body and an internal tick by an outer body.

Note: A tick associated with a narrower range of lineal bodies is a stronger statement.

1.5.06 Relative Motion of Two Non-lineal Bodies

If two non-lineal bodies q and r are moving relative to each other, how can either body know whether the other body is moving or not? We can determine the body that may not be moving as follows: Find a common outer body b of the two non-lineal bodies q and r; if either of q and r does not tick in b, then that body does not move. If both move, then we can determine which of q and r moves faster relative to the common body.

** **Figure 1.5.06**

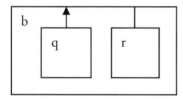

Suppose q ⊂ b, r ⊂ b, and q and r are non-lineal. Since q ticks in b, therefore, q and b have relative motion and therefore q moves. Since r does not tick in b, therefore, r and b do not have relative motion and therefore r does not move.

Assume here: q and r are two trains and b is the railway station. The train q is moving and the train r is stationary. But the passengers in either train cannot decide which train is moving by looking constantly at the other train. The passengers in a train can easily decide as to which train is moving by looking at the railway station.

Note: A common outer body for two non-lineal bodies always exists—one such body is the U body.

1.5.07 Inherited and Independent Ticks

We will call two non-empty ticks of a body b **independent ticks** if **neither** is inherited from the other; this is equivalent to: No p-tick of either tick is inherited from any p-tick of the other tick.

Two non-concurrent ticks are always independent. In Figure 1.5.07: We assume all ticks in the diagram are concurrent. Ticks T1 and T2 are non-empty; Tick T3 is empty. Also, T12 = T1 + T2; T23 = T2 + T3 = T2; T123 = T1 + T2 + T3 = T12.

Following pairs of ticks are independent: (T1, T2); (T1, T12); (T1, T23); (T1, T123); (T2, T12); (T2, T123)

Since T23 is inherited from T2 and T123 is inherited from T12, therefore, the following pairs of ticks are not independent: (T2, T23); (T12, T123).

Note: Two non-concurrent ticks are always independent.

** Figure 1.5.07

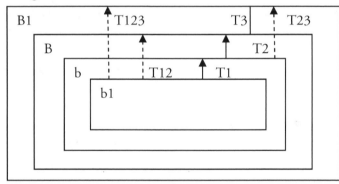

Properties of an Inherited Tick

Note: This example is not related to Figure 1.5.07.

If a tick T1 is an inherited tick entirely from a tick T2, then

- T1 and T2 are non-empty.
- T1 and T2 are concurrent.
- T1 and T2 have the same number of prime ticks.
- T1 and T2 have different object bodies and/or different reference bodies.
- T1 and T2 are logically equal.

- T1 and T2 can be reduced to the same tick by trimming empty ticks in their beginnings and/or their ends. Example: Suppose b2 ⊂ b1 ⊂ b ⊂ B ⊂ B1; T1= t[b2, b1], T2 = [b1, b], T3 = [b, B] and T4 =[B, B1]; T1 and T4 are empty ticks. Then T1 + T2 + T3 = T2 + T3 + T4. If in the equation, we trim T1 from the first tick and T4 from the second tick, then we have T2 + T3 = T2 + T3.

Note: Trivially, we could regard a tick t being inherited from t. However, we want to exclude this case from being an inherited tick as this will make every tick an inherited tick.

1.5.08 Property #3 of Ticks:
Strong Transitivity of Ticks

Suppose b1 ⊂ b ⊂ B. Transitivity of ticks implies: **If (b1 ticks in b) and (b ticks in B), then b1 ticks in B.** In view of the inheritance of ticks, we can replace "and" by "or" in the preceding transitivity statement. That is, **If (b1 ticks in b) or (b ticks in B), then b1 ticks in B**. We will call the modified transitivity statement a **strong transitivity of ticks**. In fact, the reverse of the statement is true, too. That is **if b1 ticks in B, then either (b1 ticks in b) or (b ticks in B)** since if neither b1 ticks in b nor does b tick in B, then b1 cannot tick in B either. Therefore, the "strong transitivity of ticks" statement is: **Suppose b1 ⊂ b ⊂ B. If (b1 ticks in b) or (b ticks in B), then b1 ticks in B and vice versa.**

Another Interpretation of Strong Transitivity of Ticks

Suppose b1 ⊂ b ⊂ B. If one of t[b1, b], t[b, B] or t[b1, B] is non-empty, then at least one more of the three ticks is non-empty. We can verify this as follows:

- If t[b1, b] is non-empty, then t[b1, B] is non-empty, due to inheritance of internal ticks.
- If t[b, B] is non-empty, then t[b1, B] is non-empty, due to inheritance of external ticks.
- If t[b1, B] is non-empty, then t[b1, b] is non-empty or t[b, B] is non-empty due to: t[b1, B] = t[b1, b] + t[b, B].

An Example of Strong Transitivity: Consider the following three bodies: Passenger ⊂ Train ⊂ Earth and three possible ticks: [Passenger, Train], [Train, Earth] and [Passenger, Earth]. If the passenger moves inside the train and/or the train moves relative to the Earth, then the passenger moves relative to the Earth. If the passenger moves relative to the Earth, then either the passenger moves inside the train and/or the train moves relative to the Earth. If one of the three ticks is non-empty, then at least one more of the three ticks is non-empty.

Question: Suppose train ticks relative to the Earth and passenger ticks inside the train and the two ticks are of the same length but in the opposite direction; does the passenger still tick relative to the Earth? **Yes. Here, the motion is not in terms of distance traveled; it is in terms of the number of ticks made;** in this case, the passenger makes two independent ticks.

** Figure 1.5.08

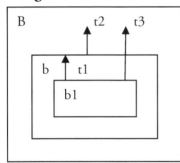

Strong Transitivity of Ticks: t3 = t1 + t2. If any one of the three ticks is non-empty then at least one more tick is non-empty.

1.5.09 Property #4 of Ticks:
Extended Prime Ticks (XP-Ticks)

Internal xp-tick: An internal prime tick of a body b or an inner body of a body b will be called an **internal extended prime tick or internal xp-tick of b.** (An internal xp-tick of a body b is different from an internal chained tick of b. An internal xp-tick of a body b is either an internal prime tick of b or any one of the prime ticks in an internal chained tick of b.)

External xp-tick of b: An external prime tick of a body b or an outer body of a body b will be called an **external extended prime tick or external xp-tick of b**. (An external xp-tick of a body b is different from an external chained tick of b. An external xp-tick of b is either an external prime tick of b or any of the prime ticks in an external chained tick of b.)

Property 4I: If b1 ⊂ b, then an internal xp-tick of a body b1 is either an internal prime tick of b or it is concurrent with some internal prime tick of b.

Property 4E: If b1 ⊂ b, then an external xp-tick of a body b is either an external prime tick of b1 or it is concurrent with some external prime tick of b1.

The proof for both properties 4I and 4E follows from the properties inheritance of ticks.

Note: The properties of inheritance of ticks apply to the ticks in general and not just the prime ticks.

1.5.10 Single-Body-Based Ticks

Normally, we define a tick t[b1, b] using two bodies b1 and b, b1 as the object body and b as the reference body, where b1 ⊂ b. We can define a tick using only one body as follows: We will denote an external tick of b by **t[b]** and an internal tick of b by **t(b).** We already use a single body to define a composite of b by **t[(b)].** We have the following interpretation of t(b), t[b] and t[(b)]:

- **t[b]** implies that there exists a body B such that b ⊂ B and b ticks in B. The body B is not unique in this case.
- **t(b)** implies that there exists b1 such that b1 ⊂ b and b1 ticks in b. The body b1 is not unique in this case.
- **t[(b)]** implies that b has an internal tick and an external tick concurrently, that is, t[(b)] = t(b) + t[b] or t[(b)] = t(b) ∪ t[b].

Note: If body b ticks, that is, t1 = t[b], then what is the lineal range of bodies for the tick t1? It is indeterminate from the statement t1 = t[b] as such but it does exist. The tick t[b] implies that there exists a body B such that t[b, B] where b ⊂ B. Then [b, B] is the range. We can also say there exists a minimal body B0 such that b ⊂ B0 ⊂ B and t[b, B0]. This makes the statement t[b, B0] stronger than t[b, B], since t[b, B0] implies t[b, B].

1.5.11 Property #5 of Ticks: Single-Body-Based Ticks

If b1 ⊂ b and t[b1, b] is a non-empty tick (prime tick or chained tick), then

- **Property #5I**: b1 has an external prime tick.
- **Property #5E**: b has an internal prime tick.
- **Property #5C**: If t[b1, b] is a chained tick, then there exists at least one b0 such that b1 ⊂ b0 ⊂ b and b0 has a composite tick; that is, b0 has an internal tick and an external tick concurrently.

The proof of the above three properties follows from the Property #1 of ticks: a non-empty tick can be decomposed into all prime ticks; a chained tick can be decomposed into two or more prime ticks.

1.5.12 Notation t[b, U]

t[b, U] and t[b] are one and the same construct. **t[b, U] is a mathematical abstraction so the U body can be treated like any other body.** Suppose b ⊂ B.

1. b ⊂ U is always true for any body b.
2. t[b, B] implies t[b, U] because of external inheritance property, but t[b, U] or t[b] does not imply t[b, B].
3. **t[b] = t[b, U].**
4. **t[b, U] = t[b, B] + t[B, U].**
5. If b ticks, then t[b, U] is non-empty and vice versa.

6. If b does not tick, then t[b, U] is empty and vice versa.
7. t[b, U] represents the absolute motion of b at an instant of time.
8. **Phantom body**: The only valid tick notations for a phantom body p are: t[p] or t[p, U]. t[p, b] and t[b, p] are invalid.

1.5.13 Property #6 of Ticks: The Equation of a Single Tick

If b1 ⊂ b and b1 ticks, then

$$t[b1] = t[b1, b] \cup t[b] \qquad \text{where } b1 \subset b \qquad (1.5.13A)$$

Or $\quad t[b1] = t[b1, b] + t[b] \qquad \text{where } b1 \subset b \qquad (1.5.13B)$

Or $\quad t[b1, U] = t[b1, b] \cup t[b, U] \quad \text{where } b1 \subset b \qquad (1.5.13C)$

Or $\quad t[b1, U] = t[b1, b] + t[b, U] \quad \text{where } b1 \subset b \qquad (1.5.13D)$

Either t[b1, b] is true, or t[b] is true, or both are true. For a single instant, we will use both operators "∪" (union) and "+" to imply logical "and/or".

1.5.14 Native Ticks

We will call an external prime tick of a body b a **native tick** if it is **not** inherited from another tick. A native tick is a prime tick but a prime tick may or may not be a native tick. In the case of the passenger-train-Earth example: If the passenger ticks in the train, then the passenger has native tick. If the passenger does not tick in the train and the train ticks, then the passenger has a prime tick but not the native tick.

Classical motion is generally formed of native ticks.

** Table 1.5.14: Example: prime and native ticks

Passenger ⊂ Train ⊂ Earth

Earth	Train	Passenger	Comments
No tick	No tick	No Tick	1. No external ticks by any of the three bodies.
No tick	No tick	Native tick	1. Train does not move. 2. Passenger walks in the train.
No tick	Native tick	1. Inherits tick as Prime tick from the train. 2. Native tick.	1. Train moves. 2. Passenger walks in the train.
No tick	Native tick	1. Inherits tick as Prime tick. 2. No native tick.	1. Train moves. 2. Passenger does not walk in the train.
Native tick	No tick	No tick	1. Passenger moves with the inherited prime tick from the Earth. 2. Train does not move. 3. Passenger does not walk.

Note: The above example can be more complicated in reality.

1.5.15 The Difference between Motions of Photons and (Material) Elementary Particles

Photons do not get a ride; they run!
(Material) elementary particles get a ride and/or run!

Here, a "ride" means inheriting motion or ticks from an outer body; a "run" means native motion or native ticks.

Neutrinos have most native motion or native ticks: Neutrinos have both native and inherited motion/ticks. However, among material elementary particles, neutrinos appear to have most native motion or native ticks.

Motion of a photon in vacuum: A tick of a photon p, in a vacuum, is represented by t[p, U]; that is, p is not an inner body of any body except U; p ticks in no body except U. This implies that all ticks of a photon in a vacuum are native ticks; a photon does not inherit ticks from another body.

Question: What happens to a photon in media such as air and water? In media, does a photon tick at all, or is it a mix of native and non-native ticks? We are unsure.

Motion of an elementary body: A tick of an elementary particle e is represented by t[e, b] + e[b] where e $\subset$ b. Either e ticks in b as a native tick and/or it inherits ticks from its outer bodies.

Limitations on Tick Representation of a Phantom Body

Suppose p is a phantom body and b is a common body. Then t[p, b] or t[b, p] do not make sense; this is particularly so in a vacuum. The only tick representations a phantom body p can have are: t[p] or t[p, U].

Note: Single body notations are of particular help, and even necessary, in the case of phantom bodies.

1.5.16 Property #7 of Ticks:
Inheritance Properties of Pure Ticks

Inheritance of pure internal ticks: A pure internal tick of a body b1 is inherited as a pure internal tick by each outer body of b1.
Proof steps:
- Suppose b1 $\subset$ b and b1 has a pure internal tick t.
- Therefore, b1 has an internal tick.

- Therefore b inherits t as an internal tick by the property of inheritance of internal ticks by outer bodies.
- We claim t cannot be a composite tick of b as in that case b would have an external tick concurrent with t and in that case b1 would have an external tick (because of the inheritance of external tick) and in that case t would not be a pure internal tick of b1, a contradiction.
- Hence a pure internal tick is inherited by an outer body as a pure internal tick.

Inheritance of pure external ticks: A pure external tick of a body b is inherited as a pure external tick by each inner body of b.
Proof steps:
- Suppose b1 ⊂ b and b has a pure external tick t.
- Therefore b has an external tick.
- Therefore b1 inherits t as an external tick by the property of inheritance of external ticks by inner bodies.
- We claim t cannot be a composite tick of b1 as in that case b1 would have an internal tick concurrent with t and in that case b would have an internal tick and in that case t would not be a pure external tick of b, a contradiction.
- Hence a pure external tick is inherited by an outer body as a pure external prime tick.

1.5.17 Composite Ticks may not be Inherited as Composite Ticks

A composite tick of a body can be inherited by an outer body either as a composite tick or as a pure internal tick. A composite tick of a body can be inherited by an inner body either as a composite tick or as a pure external tick.

1.5.18 Inheritance Table of Ticks of Different Types

Here, we list different possibilities whether a tick inherited is a pure tick or a composite tick. We assume b ⊂ B.

** Table 1.5.18: Inheritance Table

Internal/ External Inheritance Property	B	⊃	b
External tick of an outer body:	External tick	Inherited as →	External tick
External tick of an outer body:	Pure external tick	Inherited as →	Pure external tick
External tick of an outer body:	Composite tick	Inherited as →	Pure external tick or composite tick
External tick of an outer body:	Prime tick	Inherited as →	Prime tick or chained tick
External tick of an outer body:	Chained tick length = n	Inherited as →	Chained tick length ≥ n
	b	⊂	B
Internal tick of an inner body:	Internal tick	Inherited as →	Internal tick
Internal tick of an inner body:	Pure internal tick	Inherited as →	Pure internal tick
Internal tick of an inner body:	Composite tick	Inherited as →	Pure internal tick or composite tick
Internal tick of an inner body:	Prime tick	Inherited as →	Prime tick or chained tick
Internal tick of an inner body:	Chained tick length = n	Inherited as →	Chained tick length ≥ n

Note: An internal tick cannot be inherited as an external of the same body. An external tick of cannot be inherited as an internal tick of the same body. (See Axiom of Exclusion—Property #10 of ticks.)

1.5.19 Maximal External and Internal Ticks of a Body

** **Figure: 1.5.19 (Maximal External Tick of a Body:**

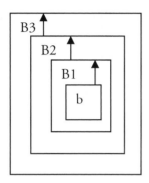

Maximal External Tick of a Body: A non-empty external tick t of a body b will be called a **maximal external tick of b** if the outermost body involved in the external tick t has no external tick.

Maximal Internal Tick of a Body: A non-empty internal tick t of a body b will be called a **maximal internal tick of b** if the innermost body involved in the internal tick t has no internal tick.

Consider the tick t[b, B3] in Figure 1.5.19. If B3 has no external tick, then t[b, B3] is a maximal external tick of b. If b has no internal tick, then tick t[b, B3] is a maximal internal tick of B3.

1.5.20 Maximal Tree of Internal XP-ticks of a Body

All independent internal xp-ticks of a body b form a tree with b as the root, nodes as inner bodies of b, branches as internal xp-ticks of b, and leaves as inner bodies of b having no internal ticks. We will call this tree as a **maximal tree of internal xp-ticks** of b.

1.5.21 Instants of Motion

We will use maximal external ticks as the **instants of motion**. We will discuss it more in Chapter 3.

1.5.22 Property #8 of Ticks: Axiom of External Ticks

1) A body cannot make two **external <u>prime</u> ticks** concurrently and independently in the same body or two **different bodies** whether lineal or non-lineal.
2) A body cannot make two **external ticks** concurrently and independently in the same body or two **<u>non-lineal</u> bodies.**

Notes:

1. A body can make two external ticks concurrently and independently in the two lineal bodies, if at least one of the two external ticks is a chained tick.
2. Two or more concurrent and independent external ticks of a body can be composed into one external chained tick. (This is proven as property #9 of ticks.)
3. A body can make two **internal ticks** concurrently and independently in the same body.

The significance of the independence of concurrent prime ticks in the above axiom: If two prime ticks are not independent, then one is inherited from the other and therefore they are concurrent.

Consequences of the Axiom of Concurrent External Ticks in different situations

A body cannot make two independent external ticks concurrently:
- If a body has two concurrent and independent external ticks then at least one of the two ticks is a chained tick. Therefore, a body cannot have two external **prime** ticks concurrently and independently. (See Figure 1.5.24E.) t1 and t4 are external

independent concurrent ticks, t1 is a prime tick but t4 is a chained tick since t4 = t1 + t2.

- A body can have two external prime ticks concurrently **if one tick is inherited from the other**. (t1 and t2 in Figure 1.5.24C and Figure 1.5.24D). Two such ticks are not independent. Therefore, this is trivial.
- A body cannot tick in two non-lineal bodies concurrently. (Figure 1.5.24G)
- A body cannot tick in an outer body repeatedly and concurrently. A special case of the above. (Fig. 1.5.24G)
- A body cannot have two non-lineal outer bodies concurrently.

A body can make two independent external ticks concurrently only if the following conditions are met:
- The reference bodies are different and lineal.
- Exactly one of the ticks is a prime tick, and other ticks are chained ticks. (See Fig 1.5.24E. t1 and t4 are two **such** ticks. Their reference bodies are lineal; only one of the two ticks, t1, is a prime tick.)

A body can make two or more independent external ticks concurrently if and only if all of the ticks are lineal. If there are two concurrent and independent external ticks of a body, then the two ticks cannot be in **parallel**; they are lineal, that is, one of the two ticks is **embedded** into the other.

Example: An external prime tick and external xp-tick of a body b can tick concurrently if the two ticks are independent.

However, concurrent internal prime ticks have no such restrictions: A body can have any number of concurrent and independent internal prime ticks. That is, two or more lineal or non-lineal inner bodies can tick in the same outer body concurrently and independently. (Figure 1.5.24F)

Suppose b2 ⊂ b1 ⊂ b and b2 does not tick in b1. Then b2 and b1 cannot tick in b concurrently and independently since t[b2, b] is inherited from [b1, b], and therefore t[b2, b] and t[b1, b] are not independent. See Figure 1.5.24H.

Significance of Concurrent External/Internal Ticks in Terms of Tick Notations:

- **Concurrent external ticks:** Suppose t1 = t[b1, b], t2 = b[b1, B]. t1 and t2 are external ticks of b1. If t1 and t2 are concurrent, then t1 and t2 must be lineal ticks and b and B must be lineal bodies.
- **Concurrent internal ticks:** Suppose t1 = t[b1, b], t2 = b[b2, b]. t1 and t2 are internal ticks of b. In this case, if t1 and t2 are concurrent, then there is no restriction. (See Figure 1.5.24B.)

Question: Suppose an electron e is shared by two atoms a1 and a2. Can e move (tick) in both a1 and a2 concurrently? Can e be an inner body of both a1 and a2 concurrently?

As per the axiom of "limitation on concurrent external ticks", the answer should be "no". However, the electron e can move (tick) in the atoms a1 and a2 alternately or intermittently with varying fineness.

The limitation imposed by the axiom of "limitation on concurrent external ticks" is localized as can be seen from Figure 1.5.24J; it is hidden from the outer bodies.

1.5.23 Property #9 of Ticks: Maximal Ticks of a Body

All concurrent independent external xp-ticks of a body b can be composed into a unique maximal external tick of b.

All concurrent independent internal xp-ticks of a body b can be composed into one or more maximal external ticks of b.

Proof

Without any loss of generality, we will prove it only for the case when body b has three concurrent and independent xp-ticks t1, t2 and t3. Let t1 = t[b, B], t2 = t[b2, B2] and t3 = t[b3, B3] where b ⊂ B, b2 ⊂ B2 and b3 ⊂ B3. In addition, by the axiom of concurrent external ticks, the bodies B, B2 and B3 must be lineal. Therefore, we can assume that b ⊂ B ⊂ b2 ⊂ B2 ⊂ b3 ⊂ B3. Therefore, t[b, B3] = t[b, B] + t[B, b2] + t[b2, B2] + t[B2, b3] + t[b3, B3]; we can assume that t[B, b2], t[B2, B3] are empty ticks since the body b has only three concurrent xp-ticks. t[B, B3] is precisely the maximal external tick.

Similarly, we can prove the other property.

** **Figure 1.5.23**

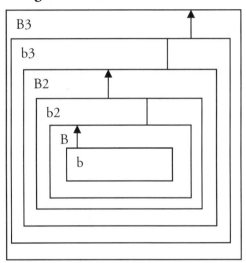

Properties of Maximal External Tick

- A body b can have at most one maximal external tick concurrently. A body b can have at most one maximal external tick concurrently. (But a body can have more than one **maximal internal tick concurrently**.)
- A body b can have at most one maximal external tick concurrently. A maximal external tick of b may be composed of 0, 1 or more prime ticks; accordingly, the maximal tick is empty, a prime tick or a chained tick.
- A body b can have at most one maximal external tick concurrently. If a maximal external tick is non-empty, then the outermost xp-tick (in the maximal tick) is a pure internal prime tick.
- A body b can have at most one maximal external tick concurrently. A body b can have at most one pure external xp-tick out of all concurrent external xp-ticks of the body.

- A body b can have at most one maximal external tick concurrently. The first xp-tick, t1, in a non-empty maximal external tick of b, is a prime tick of b and this one is a pure prime tick in case b has no internal tick concurrently with t1.

1.5.24 Property #10 of Ticks: Axiom of Exclusion

1. **Same tick cannot be both an internal tick and an external tick of the same body.**
2. **An internal tick cannot be inherited as an external tick of the same body. An external tick cannot be inherited as an internal tick of the same body.**

The above two properties of ticks will be called the **axiom of exclusion** and will be assumed without proof. The axiom implies:

3. Concurrent internal tick and external tick of the same body must be different and independent ticks (one cannot be inherited from the other).
4. An internal tick of an inner body of b cannot be inherited as an external tick of an outer body of b. An external tick of an outer body b cannot be inherited as an internal tick of an inner body of b. (This property and the axiom of inheritance are complimentary.)
5. If two ticks t1 and t2 are not independent (one is inherited from the other), then they cannot form a composite tick.
6. To form a composite of a body b, we need two concurrent and independent ticks of a body b, an internal tick of b and an external tick of b.
7. If two ticks t1 and t2 form a composite tick then the two ticks are independent and concurrent. (See Figure 1.5.24A.) Converse may not be true (See figure 1.5.24B).
8. If two ticks t1 and t2 are independent and concurrent then t1 and t2 either form a composite tick <u>or</u> they are non-lineal internal ticks of a body. (See Figure 1.5.24A and Figure 1.5.24.B.)

** **Figure 1.5.24A**   **Figure 1.5.24B**

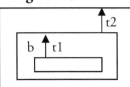
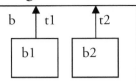

** **Figure 1.5.24D**   ** **Figure 1.5.24C**

Universe Figure 1-4	Universe Figure 1-3
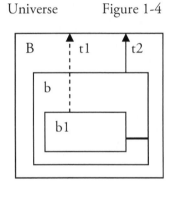	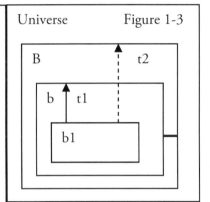
b1 inherits external tick t2 of b as external tick t1; t1 and t2 cannot form a composite tick.	B inherits internal tick t1 of b1 as internal tick t2; t1 and t2 cannot form a composite tick.

** Figure 1.5.24E

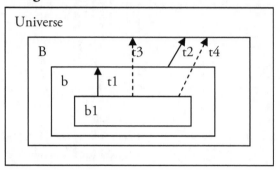

t3 = t4 = t1 + t2; each of t3 and t4 inherits
both t1 and t2; t1 is an internal tick of b and
t2 is an external tick of b.

** Figure 1.5.24G   *** Figure 1.5.24F

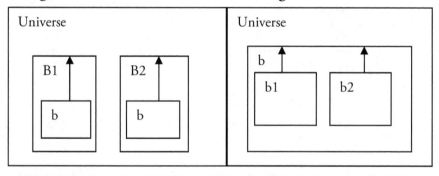

| Body b cannot have two external prime ticks concurrently and independently if B1 and B2 are non-lineal. | Body b can have two internal prime ticks concurrently and independently |

** **Figure 1.5.24H**

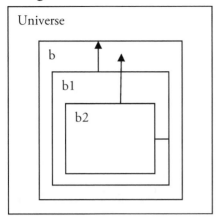

Two non-lineal bodies b1 and b2 cannot tick concurrently and independently in outer body b if b2 cannot tick in b1.

** **Figure 1.5.24J**

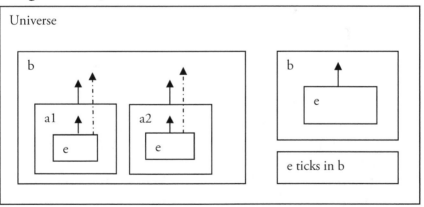

e ticks in a1 and a2 intermittently. The axiom of inheritance of internal ticks implies that e ticks in b whenever e ticks in a1 or a2. The figure on the left is a microscopic view and the figure on the right is a macroscopic view of the relation between e and b.

1.6 List of Types of Ticks

Notations:

- $b1 \subset b \subset B$, $b11 \subset b \subset B$
- $t1 = t[b1, b]$, $t11 = t[b11, b]$, $t2 = t[b, B]$
- $t3 = t1 + t2 = t[b1, B] = t[b1, b] + t[b, B]$
- $t4 = t11 + t2 = t[b11, B]$

We assume that $t1$, $t11$ and $t2$ are prime ticks.

- $t[b]$: External tick of b: There exists a body B such that $b \subset B$ and $t[b, B]$.
- $t(b)$: Internal tick of b: There exists a body b1 such that $b1 \subset b$ and $t[b1, b]$.
- $t[(b)]$: Composite tick of b: There exists bodies b1 and B such that $b1 \subset b \subset B$ and $t[b1, b]$ and $t[b, B]$.

Ticks: A tick is either an empty tick, or a prime tick, or a chained tick

- **Empty/Non-Empty Ticks**
  The tick $t[b1, b]$ is empty if $b1 \subset b$ but b1 does not tick in b. If a tick is not empty, then it is non-empty.
- **External Ticks**
  t1 is an external of b1 in b.
  t2 is an external tick of b in B.
  t3 is an external tick of b1 in B.
- **Internal Ticks**
  t1 is an internal tick of b.
  t2 is an internal tick of B in the case $t[b, B]$.
  t3 is an internal tick of B in the case $t[b1, B]$.
- **Prime Ticks: non-empty and not chained ticks.**
  t1 and t2 are prime ticks
  o **External Prime Ticks: t[b1, b], t[b1]**
    t1 is an external prime tick of b1 in b.
    t2 is an external prime tick of b in B.
  o **Internal Prime Ticks: t[b1, b], t(b)**
    t1 is an internal prime tick of b.
    t2 is an internal prime tick of B.
- **Native Ticks**
  External prime ticks not inherited from other ticks

- **Chained Ticks**
  - o **External Chained Ticks: t[b1, B], t[b1]**
    t3 is an external chained tick of b1 in B.
  - o **Internal Chained Ticks: t[b1, B], t[b]**
    t3 is an internal chained tick of B.
- **Composite Ticks: t3 = [b1, b] + t[b, B] or t[(b)]**
  t3 is a composite tick of b
- **Lineal Ticks:** t1 and t2 are lineal ticks.

** **Figure 1.6**

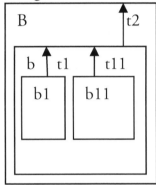

- **Non-lineal Ticks**
  Ticks t1 and t11 are non-lineal.
- **Extended Prime Ticks (xp-ticks)**
  - o **Internal xp-tick**
    t1 is an internal xp-tick of B.
  - o **External xp-tick**
  - o t2 is an external xp-tick of b.
- **Maximal External Ticks**
  All concurrent external independent prime ticks of both a body b
  and outer bodies of b can be composed into a single external tick t
  of b. The tick t is called a **maximal external tick** of the body b.
  - o t3 is the maximal external tick of b1.
  - o t4 is the maximal external tick of b11.
- **Independent Ticks**
  Two ticks are independent if neither is inherited from the other.
  Ticks t1 and t2 are independent external ticks of b1.
  Ticks t1 and t11 are independent internal ticks of b.

1.7 A General Diagram Demonstrating Most Types of Ticks and Their Properties

** **Figure 1.7**

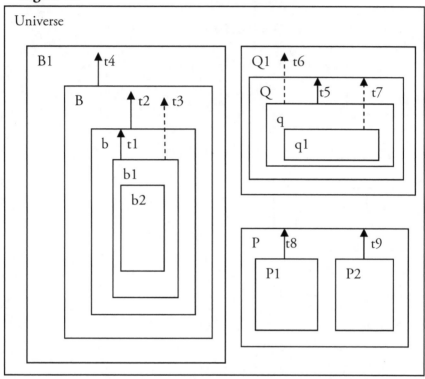

General Diagram to demonstrate most types of ticks and their properties:

1.7.1 Types of Ticks in the general diagram (Figure 1.7)

Empty Ticks
t[b2, b1]
Non-empty Ticks
All named ticks

Prime Ticks (p-ticks)

t1 = t[b1, b], t2 = [b, B], t4 = [B, B1]
t5 = t[q, Q], t8 = t[p1, P], t9=[p2, P]

Internal Ticks

t1 = t(b), t2 = (B), t4 = (B1)
t5 = t(Q), t8 = t(P), t9=(P)

External Ticks

t1 = t[b1], t2 = [b], t4 = [B]
t5 = t[q], t8 = t[p1], t9=[p2]

External Extended Prime Ticks (xp-ticks)

t1, t2 and t4 are external xp-ticks of b1.
t2 and t4 are external xp-tick of b.

Internal Extended Prime Ticks (xp-ticks)

t1, t2 and t4 are internal xp-ticks of B1.
t1 and t2 are internal xp-tick of B.
Note: All prime ticks are xp-ticks, too.

Pure Internal Ticks

t4 = t((B1)), t5 = ((Q)), t6 = t((Q1)), t7 = t((Q)),
t8 = t((P)), t9 = t((P))

Pure External Ticks

t1=t[[b1]], t3=t[[b1]], t5=t[[q]], t8=[[p1]], t9 = [[p2]]

Chained Ticks

t3 = t[b1, B] = t[b1, b] + t[b, B] = t1 + t2
t1 + t2 + t4 = t[b1, b] + t[b, B] + t[B, B1] = t[b, B1]

Composite Ticks

t[(b)], t[(B)]

Inherited Ticks

t1 an internal tick of b is inherited as t3 an internal tick of B.
t2 an external tick of b is inherited as t3 an external tick of b1.
Note: In the above two cases, inherited ticks result in composite ticks.
t5 an internal tick of Q is inherited as t6 an internal tick of Q1.
t5 an external tick of q is inherited as t7 an external tick of q1.

Independent Ticks
t5 and t6 are not independent.
t5 and t7 are not independent.
t6 and t7 are not independent.
t1 and t2 are independent.
t1 and t3 are independent.
t2 and t3 are independent.
However, t1, t2 and t3 are not all independent.
Note: This is because t3 is a chained tick, composed of two independent prime ticks t1 and t2. It is not so in case of t5, t6 and t7.
t1, t2 and t4 are independent.

Concurrent and Independent Internal Prime Ticks
T8 and t9 of P are Concurrent and Independent Internal Prime Ticks

Concurrent and Independent External Prime Ticks
(No two concurrent and independent external prime ticks of any body are possible, as mandated by the axiom of limitation on concurrent and independent external ticks.)
t1 and t3 are concurrent and independent external ticks of b1 where t1 is a prime tick and t3 is a chained tick

Maximal External Tick
t1 + t2 + t4 (= t[b1, B1]) is a **maximal external tick** of b1.

CHAPTER 2

Theory of Concurrency of Events

2.01 Introduction

We introduced ticks in Chapter 1 and analyzed them extensively. A tick is also an event. Each event has the following important attribute explicitly or implicitly: an instant of time. In Chapter 1, we discussed cases of ticks that occurred primarily at a single instant of time. For examples: a prime tick involving a single tick or a chained tick involving two or more concurrent ticks. We will now also use sets of ticks where ticks in a set occur at one or more instants of time. In later chapters, we will show that the concurrency of ticks plays a pivotal role in anything involving ticks.

Our goal in this chapter is to develop a theory of concurrency of events, which is generalized enough and with focus on ticks as events:

- To determine concurrency of ticks as events.
- To understand and gain insight into concurrency of ticks.
- To prevent pitfalls in the logic of concurrency of events when applying it to ticks.

2.02 Concurrency of Events and Order of Events

Concurrency of Events

What is concurrency of two events?

Time can be viewed as the order of the occurrences of events.
Suppose, E1 and E2 are two events. There are three possibilities:
  E1 occurs earlier than E2.
  E1 occurs later than E2.
  E1 and E2 occur concurrently (simultaneously).

Thus a need for concurrency of two events arises when we try to order the occurrences of events.

We will assume concurrency of events is a primitive concept; we will not define it. Generally, as the number (density) of events increases, the likelihood of concurrency of events also increases.

Order of Events

Notation for the order of two events: Suppose x and y are two events. We will denote the relation "x occurs earlier than y" by $x \rightarrow y$ and the relation "x occurs later than y" by $x \leftarrow y$.

Neither of the two relations $x \rightarrow y$ and $x \leftarrow y$ is reflexive or symmetric; both relations are transitive.

Notation for concurrency of two ticks: If two ticks x and y are concurrent, then we will denote the relation "x is concurrent with y" by $x \leftrightarrow y$.

  It is reflexive: That is, $x \leftrightarrow x$.
  It is symmetric: That is, if $x \leftrightarrow y$, then $y \leftrightarrow x$.
  It is transitive: That is, if $x \leftrightarrow y$ and $y \leftrightarrow z$, then $x \leftrightarrow z$.

Note: This theory of concurrency of events may not meet the rigors of experimental physics where two events may not be in the same order for two different observers. Here, we propose a mathematical theory of concurrency of events independent of observers.

2.03 Concurrency Disjoint (C-Disjoint) Sets of Events

Definition: Set of Events (E-Set). An e-set is a set of events.

Definition: We will call two sets of events **c-disjoint** if no member from one set is concurrent with any member of the other set.

Question: Is an event concurrent with itself? This issue arises in the following context: Suppose P and Q are two sets of events having a common event x, that is, $x \in P$ and $x \in Q$. Therefore, P and Q are not disjoint. Can P and Q be c-disjoint if x belongs to both P and Q?

The answer will depend on whether x of P is deemed to be concurrent with x of Q or not. **We will assume x is concurrent with x. However, if x and y are concurrent, then it does not imply that x and y are the same event.**

Concurrency Equivalent (C-Equivalent) E-Sets

Definition: We will call two e-sets P and Q **concurrency equivalent (or c-equivalent)** if a member of **either** e-set is concurrent with some member of the other e-set.

The concurrency equivalence relation is reflexive, symmetric and transitive. This is because the relation $x \leftrightarrow y$ is reflexive, symmetric and transitive.

Note: The **"either"** word is important in the above definition. It is possible that each member of P is concurrent with some member of Q but each member of Q is not concurrent with some member of P. For example: Suppose P = {e11, e22} and Q = {f11, f12, f13} and e11$\leftrightarrow$ f11, e22 $\leftrightarrow$ f12. In this case, each member of P is concurrent with some member of Q but f13 of Q is not concurrent with any member of P. Therefore, P and Q are not c-equivalent.

2.04 Set of Prime Ticks (Prime Hulchul or P-Hulchul)

Since we will apply the theory of concurrency of events to ticks, we will now introduce some of the specific sets of ticks we will actually use in later chapters. We will call a set of prime ticks of a body b a **prime hulchul** or **p-hulchul of the body b**.

Internal P-Hulchul of a Body
ph(b) = Set of all independent internal prime ticks of a body b.

External P-Hulchul of a Body
ph[b] = Set of all independent external prime ticks of a body b.

Relative P-Hulchul of a Body
ph[b, B] = Set of all independent external prime ticks of a body b in any body b0 such that b ⊂ b0 ⊆ B.

Pure Internal P-Hulchul of a Body
ph((b)) = Set of pure internal prime ticks of b
$\qquad$ = {x | x ∈ ph(b) and not (x ↔ y) for any y ∈ ph[b]}.

Pure External P-Hulchul of a Body
ph[[b]] = Set of pure external prime ticks of b
$\qquad$ = {x | x ∈ ph[b] and not(x ↔ y) for any y ∈ ph(b)}.

Composite Internal P-Hulchul of a Body
ph[I(b)] = Set of composite internal p-ticks of b
$\qquad$ = {x | x ∈ ph(b) and x ↔ y for some y ∈ ph[b]}.

Composite External P-hulchul of a Body
ph[E(b)] = Set of composite external p-ticks of b
$\qquad$ = {x | x ∈ ph[b] and x ↔ y for some y ∈ ph(b)}.

Total P-hulchul of a Body
ph{b} = ph(b) ∪ ph[b]
$\qquad$ = Set of all independent (internal or external) p-ticks of b.

The difference between the p-hulchuls ph[b] and ph[b, B] is as follows: In the case of ph[b], b may tick in any body b0 such that $b \subset b0$. In the case of ph[b, B], b may tick only in bodies b0 such that $b \subset b0 \subseteq B$. Thus $ph[b, B] \subseteq ph[b]$.

2.05 Properties of the Sets of Prime Ticks (P-Hulchuls)

1. The flowing pairs of sets are disjoint either by their definitions or because of the axiom of exclusion of internal and external ticks of b:

ph(b) and ph[b]	by axiom of exclusion
ph((b)) and ph[[b]]	by axiom of exclusion
ph((b)) and ph[E(b)]	by axiom of exclusion
	Since $ph((b)) \subseteq ph(b)$, $ph[E(b)] \subseteq ph[b]$
ph[[b]] and ph[I(b)]	by axiom of exclusion
	Since $ph[[b]] \subseteq ph[b]$, $ph[I(b)] \subseteq ph(b)$
ph[I(b)] and ph[E(b)]	by axiom of exclusion
	Since $ph[I(b)] \subseteq ph(b)$, $ph[E(b)] \subseteq ph[b]$
ph((b)) and ph[I(b)]	by their definition
	$ph((b))$ not($\leftrightarrow$) $ph[[b]]$ and
	$ph[I(b)] \leftrightarrow ph[[b]]$
ph[[b]] and ph[E(b)]	by their definition
	$ph[[b]]$ not($\leftrightarrow$) $ph((b))$ and
	$ph[E(b)] \leftrightarrow ph((b))$

1. **$ph[b, B] \subseteq ph(b)$**
    as explained earlier.

2. **$ph((b)) \subseteq ph(b)$**
    $ph[I(b)] \subseteq ph(b)$
    $ph(b) = ph((b)) \cup ph[I(b)]$

    $ph[[b]] \subseteq ph[b]$
    $ph[E(b)] \subseteq ph[b]$
    $ph[b] = ph[[b]] \cup ph[E(b)]$
    by their definitions

3. **ph(b1) ⊆ ph(b) where b1 ⊂ b.**
   That is, ph(b) remains the same or increases outer-body-ward.
   This is due to inheritance of internal ticks of a body b1 by an outer
   body b of b1.

4. **ph[b] ⊆ ph[b1] where b1 ⊂ b.**
   That is, ph[b] remains the same or decreases outer-body-ward.
   This is due to inheritance of external ticks of a body b by an inner
   body b1 of b.

5. **ph((b1)) ⊆ ph((b)) where b1 ⊂ b.**
   That is, ph((b)) increases or remains the same outer-body-ward.
   This is due to inheritance of pure internal ticks of a body b1 by an
   outer body of b1 by b.

6. **ph[[b]] ⊆ ph[[b1]] where b1 ⊂ b.**
   That is, ph[[b]] decreases or remains the same outer-body-ward.
   This is due to inheritance of pure external ticks of a body b by an
   inner body of b1.

7. **ph[I(b)] and ph[E(b)] have no inheritance property similar to
   those of ph(b), ph[b], ph((b)) and ph[[b]].**

8. **ph{b} = ph((b)) ∪ ph[I(b)] U ph[[b]] U ph[E(b)]**

*Since ph(b) increases and ph[b] decreases outer-body-ward, is it possible
that their sum, in some manner, is the same for all bodies? Actually it is
true, if ph(b) and ph[b] are measured in terms of their concurrency. We
will prove it in Chapter 3.*

2.06 Events and Sets of Events
(E-Sets)—Assumptions

1. We will not define an event; we assume what it is. A tick is an
   example of an event. We will use sets of events a lot. We will
   call a set of events an **e-set**.

2. **Concurrency ID (CID)**: Each event has the following attribute explicitly or implicitly: a definite instant of time—it is like a timestamp on each event. We will call the attribute a **Concurrency Id or CID**.

   Two events are concurrent if and only if the CID's of the two events are the same.

   Note: At this stage of our research, we do not have any theoretical mechanism available to us to determine an explicit CID of an event. However, we will assume CID's do exist in principle.

   We will use CID both as a function and as an object. For example: If x = CID(X), where X is an event, then x is a CID.

3. **An e-set is a sequence**: A set of events is, in fact, a sequence of events in the order of their occurrences.

4. **Any two events can be compared theoretically to determine whether or not they are concurrent.**

5. **An e-set itself may be an event**: A set of events, with all events occurring at the same instant, is also an event because it has a unique CID. On the other hand, a set of events, not all of them occurring at the same instant, is not an event.

6. **P-hulchuls are e-sets**: P-hulchuls ph(b), ph[b], ph[b, B], ph((b)), ph[[b]], ph[I(b)], ph[E(b)], and ph{b} all are sets of events.

7. **Union, intersection, and difference of two e-sets**: If P and Q are two sets of events, then their union, intersection, and difference are also sets of events.

8. **Examples of objects that are not events**: A number is not an event; a hobby is not an event. (An event usually implies an action.)

9. **Concurrency of an event with some member of an e-set**: If an event x is concurrent with some member of e-set P and P ⊂ Q, then x is concurrent with some member of Q, too. If x is not concurrent with any member of Q, then x is not concurrent with any member of P, either.

2.07 Concurrency Sets (C-Sets)

Definition: We will call a set S of events a **concurrency set or c-set** if no two members of S are concurrent.

This definition implies that each member in a c-set has a unique CID.

2.08 Set of Concurrency IDs

A set of CID's will be denoted by **CIDS(P).** CIDS(P) exists only if P is a c-set because a set cannot have two identical members. We will enclose a CID in CIDS(S) by a pair of angular brackets just to imply the presence of a CID.

Example: Suppose P = {e11, e12, e13, e14} and Q = {f11, f21, f13} are two sets of events. We assume that the last character of an event represents the CID of the event. No two events in S are concurrent since CID's of no two events in P are the same. CIDS(P) = {<1>,<2>,<3>,<4>}. Therefore, P is a c-set. In the case of Q, f11 and f21 are concurrent since both have same CID "<1>". Therefore, Q is not a c-set and CIDS(Q) does not exist in this case.

2.09 Difference between E-Sets and C-Sets

Members of any two e-sets may belong to two different classes, whereas members of all c-sets belong to the same class—the class of CID's. Therefore, members of all c-sets, all being CID's, are comparable.

2.10 Properties of C-Sets and C-Disjoint Sets

Note: A c-set is also a set. Therefore, a c-set has all the properties of a set. Members of a c-set have unique CID's. Members of a c-set can be represented either by unique CID's only or by any unique names with unique CID's.

1. **A subset of a c-set is a c-set.**
    Suppose S1 is a subset of S. Because no two members of S1 can be concurrent as that would imply those two members are concurrent in S as well, a contradiction of the fact that S is a c-set and a c-set cannot have two concurrent members.

2. **Intersection of two c-sets is a c-set.**
    This is because intersection of two c-sets is a subset of either of the two c-sets.

3. **Difference of two c-sets is a c-set**
    This is because the difference of two c-sets is a subset of one of the two c-sets.

4. **Union of two c-sets is a c-set (even if the two c-sets are not c-disjoint based on the assumption each c-set is a set of distinct CID's).**

    Union of two c-sets is like a union of two sets; union of two sets is the set of all distinct members of the two sets. Therefore, union of two c-sets contains distinct CID's. Therefore, union of two c-sets is a c-set.

    Example:
    P = {<1>,<2>}, Q = {<1>,<3>,<4>}.
    P and Q are c-sets and they are not c-disjoint.
    P ∪ Q = {<1>,<2>,<3>,<4>}; it is a c-set.
    P ∩ Q = {<1>}.
    P – Q = {<2>}.
    Q – P = {<3>,<4>}.

5. **Two subsets of a c-set are c-disjoint.**

    Suppose P1 and P2 are subsets of a c-set P. Then each of P1 and P2 must be a c-set. In addition, P1 and P2 must be c-disjoint because if P1 and P2 are not c-disjoint, then there exists an x in P1 and a y in P2 such that x and y are concurrent. Since x and y also belong to P, therefore, P is not a c-set, a contradiction of our assumption that P is a c-set.

6. **If two sets of events P and Q are c-disjoint, P1 ⊆ P and Q1⊆ Q, then P1 and Q1 are c-disjoint.**

    Suppose P1 and Q1 are not c-disjoint. Then there exist x and y such that x ∈ P1 and y ∈ Q1 and x ↔ y. Therefore, x ∈ P and x ∈ Q and x ↔ y. Therefore, P and Q are not c-disjoint, a contradiction.

    Note: Here, either of P and Q may or may not be a c-set.

7. **If P and Q are c-sets and P and Q are not c-disjoint, then a member of either c-set may be concurrent with at most one member of the other c-set.**

    If possible, let x ↔ y1 and x ↔ y2 where x ∈ P, y1 ∈ Q and y2 ∈ Q. Therefore, using the properties that concurrency of two events is symmetric and transitive, y1 ↔ y2; this contradicts the assumption that Q is a c-set.

8. **Suppose P and Q sets of events. If P and Q are c-disjoint, then P and Q are disjoint.**

    If P and Q are not disjoint, then there should exist an x such that x ∈ P and x ∈ Q. Since x is the same event in the sets P and Q, therefore, x of P is concurrent with x of Q contradicting the assumption that P and Q are c-disjoint.

    Note: However, if P and Q are disjoint, then P and Q may or may not be c-disjoint.

2.11 Concurrency Operator CNCY

We will now introduce an operator, we will call **Concurrency Operator CNCY,** to define concurrency for any set of events and derive a unique c-set from the given set of events.

Definition: Suppose S is a set of events. We will define the **Concurrency Operator CNCY** as follows:

Divide all events in S into classes of concurrent events. This implies that all events in a class are concurrent but no two events, chosen one each from the two different classes, are concurrent. We will denote the set of the classes of concurrent events of S by **CNCY(S);** we will call members of CNCY(S) as **concurrency events or c-events.** Members of CNCY(S) are also events.

Question: Is the division of S into the classes (c-events) of concurrent events unique? **Yes.**

Question: What are the characteristics of the contents of CNCY(S) like?

CNCY(S) is a c-set. Its members are distinct CID's of the members of S.

2.12 Concurrency Difference Operator CDIF

We will now create concurrency version of the difference of two sets. If P and Q are two sets, then their difference is defined as follows:

$P - Q = \{x \mid x \in P \text{ and } x \notin Q\}$.

CDIF(P, Q) = {x | x ∈ P and not (x ↔ y) for any y ∈ Q} where P and Q are e-sets

Properties of CDIF:
1. $CDIF(P, Q) \subseteq P$.
2. $CDIF(Q, P) \subseteq Q$.

3. In general, CDIF(P, Q) and CDIF(Q, P) are different.
4. We can define: **ph((b)) = CDIF(ph(b), ph[b])**
5. We can define: **ph[[b]] = CDIF(ph[b], ph(b))**

2.13 Concurrency Intersection Operator CINT

We will now create a concurrency version of the intersection of two sets. If P and Q are two sets, then their intersection is defined as follows: P ∩ Q = {x | x ∈ P and x ∈ Q}. Whereas P ∩ Q = Q ∩ P, CINT(P, Q) and CINT(Q, P) are different in general.

CINT(P, Q) = {x | x ∈ P and x ↔ y for some y ∈ Q} where P and Q are e-sets

Properties of CINT:

1. CINT(P, Q) ⊆ P
2. CINT(Q, P) ⊆ Q
3. In general, CINT(P, Q) and CINT(Q, P) are different. However, We will prove: **CNCY(CINT(P, Q)) = CNCY(CINT(Q, P))** (Property #11 of Concurrency Operator)
4. We can now define: **ph[I(b)] = CINT(ph(b), ph[b])**
5. We can now define: **ph[E(b)] = CINT(ph[b], ph(b))**

2.14 Properties of the Concurrency Operator CNCY

1. **Output of the operator CNCY is a c-set.**

   This is by the definition of the operator CNCY.

2. **CNCY(P) = P if P is a c-set.**

3. **CNCY(CNCY(P)) = CNCY(P).**

   This is because CNCY(P) is a c-set. This means the concurrency operator CNCY is **idempotent**.

4. **If P and Q are c-disjoint sets of events, then CNCY(P) and CNCY(Q) are also c-disjoint.**

   Proof: Suppose P and Q are c-disjoint. If CNCY(P) and CNCY(Q) are not c-disjoint, then CNCY(P) and CNCY(Q) have a common member, which is a CID. This CID must be the same for some member of P and some member of Q implying that P and Q are not c-disjoint, a contradiction of our assumption. Therefore, CNCY(P) and CNCY(Q) are c-disjoint.

5. **Even if e-sets P and Q are not equal, CNCY(P) and CNCY(Q) may be equal.**

   For example, suppose P = {e11, e22} and Q = {f11, f21, f32}. Therefore, CNCY(P) = {<1>, <2> and CNCY(Q) = {<1>,<2>}. Thus P and Q are not equal but CNCY(P) and CNCY(Q) are equal.

6. **If P ⊂ Q, then CNCY(P) ⊆ CNCY(Q) where P and Q are e-sets.**

   Proof: Since P ⊂ Q, therefore, Q has additional member(s) that may or may not belong to any concurrency classes of P. This may result in additional concurrency classes for Q.

   Note: It is possible that CNCY(P) = CNCY(Q) even if P ⊂ Q. For example: If P = {e11, e22} and Q = {e11, e22, e32}, then CNCY(P) = CNCY(Q) = {<1>,<2>}.

7. **Concurrency Union of Two E-SETS:**
   CNCY(P ∪ Q) = CNCY(P) ∪ CNCY(Q) where P and Q are e-sets that may or may not be disjoint or c-disjoint.
   Proof Steps:
   - Suppose P and Q are e-sets.
   - **First, we will prove: CNCY(P ∪ Q) ⊆ (CNCY(P) ∪ CNCY(Q)).**
   - Let x ∈ CNCY(P ∪ Q).
   - Therefore, x = CID(X) for some X ∈ (P ∪ Q).
   - Therefore, X ∈ P or X ∈ Q.
   - Therefore, x ∈ CNCY(P) or x ∈ CNCY(Q).
   - Therefore, x ∈ (CNCY(P) ∪ CNCY(Q)).

- Therefore, CNCY(P ∪ Q) ⊆ (CNCY(P) ∪ CNCY(Q)).
- **Now we will prove: (CNCY(P) ∪ CNCY(Q)) ⊆ CNCY(P ∪ Q).**
- Let x ∈ (CNCY(P) ∪ CNCY(Q)).
- Therefore, x ∈ CNCY(P) or x ∈ CNCY(Q).
- Therefore, x = CID(X) for some X ∈ P or x = CID(Y) for some Y ∈ Q.
- Therefore, X ∈ (P ∪ Q) or Y ∈ (P ∪ Q).
- Therefore, x ∈ CNCY(P ∪ Q)
- Therefore, (CNCY(P) ∪ CNCY(Q)) ⊆ CNCY(P ∪ Q).
- **Therefore, CNCY(P ∪ Q) = (CNCY(P) ∪ CNCY(Q)).**

Using the above property for p-hulchuls:
P = ph(b) and Q = ph[b]
CNCY(ph(b) ∪ ph[b]) = CNCY(ph(b)) ∪ CNCY(ph[b])

8. Concurrency Difference of Two E-Sets
CNCY(CDIF(P, Q)) = CDIF(CNCY(P), CNCY(Q)) where P and Q are e-sets:

Proof Steps:
- **First we will prove: CNCY(CDIF(P, Q)) ⊆ CDIF(CNCY(P), CNCY(Q)).**
- **Let x** ∈ CNCY(CDIF(P, Q)).
- Therefore, x = CID(X) for some X ∈ CDIF(P, Q).
- Therefore, X ∈ P and not (X ↔ Y) for any Y ∈ Q.
- Therefore, x ∉ Q.
- Therefore, x ∈ CNCY(P) and x ∉ CNCY(Q).
- Therefore, x ∈ CDIF(CNCY(P), CNCY(Q)).
- **Now we will prove: CDIF(CNCY(P), CNCY(Q)) ⊆ CNCY(CDIF(P, Q)).**
- **Let x** ∈ CDIF(CNCY(P), CNCY(Q)).
- We note that CNCY(P) and CNCY(Q) are c-sets.
- Therefore, x ∈ CNCY(P) and x ∉ CNCY(Q).
- Therefore, x = CID(X) for some for some X ∈ P and x ≠ CID(Y) for any Y ∈ Q.
- Therefore, X of P is not concurrent with any Y of Q.
- Therefore, X ∈ CDIF(P, Q).

- Since x = CID(X), therefore, x ∈ CNCY(CDIF(P, Q)).
- **Therefore,**
  CDIF(CNCY(P), CNCY(Q)) ⊆ CNCY(CDIF(P, Q)).
- **Therefore,**
  CNCY(CDIF(P, Q)) = CDIF(CNCY(P), CNCY(Q)).

Using the above property for p-hulchuls:
P = ph(b) and Q = ph[b]
CNCY(CDIF(ph(b), ph[b])) = CDIF(CNCY(ph(b)),
  CNCY(ph[b])).
Or
CNCY(ph((b))) = CDIF(CNCY(ph(b)), CNCY(ph[b])).
Similarly,
CNCY(ph[[b]]) = CDIF(CNCY(ph[b])), CNCY(ph(b)).

9. **CNCY(CINT(P, Q)) = CINT(CNCY(P), CNCY(Q)) where**
   P and Q are e-sets.
   Proof Steps:
   - **First, we will prove: CNCY(CINT(P, Q)) ⊆**
     CINT(CNCY(P), CNCY(Q)).
   - Let x ∈ CNCY(CINT(P, Q)).
   - Therefore, x = CID(X) for some X ∈ (CINT(P, Q)).
   - Therefore, X ∈ P and X ↔ Y for some Y ∈ Q.
   - Therefore, x = CID(X) where X ∈ P.
   - Therefore, x ∈ CNCY(P).
   - Since x = CID(X) and X ↔ Y for some Y ∈ Q, therefore x
     = CID(Y) for some Y ∈ Q.
   - Therefore, x ∈ CNCY(Q).
   - Therefore, x ∈ CINT(CNCY(P), CNCY(Q)).
   - Therefore, CNCY(CINT(P, Q)) ⊆ CINT(CNCY(P),
     CNCY(Q)).
   - **Now we will prove: CINT(CNCY(P), CNCY(Q)) ⊆**
     CNCY(CINT(P, Q)).
   - Let x ∈ CINT(CNCY(P), CNCY(Q)).
   - Therefore, x ∈ CNCY(P) and x ∈ CNCY(Q).
   - Therefore, x = CID(X) for some X ∈ P and x = CID(Y) for
     some Y ∈ Q. Therefore, X ↔ Y.
   - Therefore, X ∈ (CINT(P, Q)).

- Since x = CID(X), therefore, x $\in$ CNCY((CINT(P, Q)).
- Therefore,
  CINT(CNCY(P), CNCY(Q)) $\subseteq$ CNCY(CINT(P, Q)).
- Therefore,
  CNCY(CINT(P, Q)) = CINT(CNCY(P), CNCY(Q)).

Using the property for p-hulchuls:
P = ph(b) and Q = ph[b]:
CNCY(CINT(ph(b), ph[b])) = CINT(CNCY(ph(b)),
   CNCY(ph[b])).
Or
CNCY(ph[I(b)] = CINT(CNCY(ph(b)), CNCY(ph[b])).
Similarly,
CNCY(ph[E(b)] = CINT(CNCY(ph[b]), CNCY(ph(b))).

10. CNCY(CINT(P, Q)) = CNCY(CINT(Q, P))
Proof Steps:
- Let x $\in$ CNCY(CINT(P, Q)).
- Therefore, x = CID(X) for some X $\in$ CINT(P, Q).
- Therefore, X $\in$ P and X $\leftrightarrow$ Y for some Y $\in$ Q.
- Therefore, x = CID(Y).
- Therefore, x $\in$ CNCY(CINT(Q, P)).
- Therefore, CNCY(CINT(P, Q)) $\subseteq$ CNCY(CINT(Q, P)).
- Similarly, we can prove:
  CNCY(CINT(Q, P)) $\subseteq$ CNCY(CINT(P, Q)).
- Therefore, **CNCY(CINT(P, Q)) = CNCY(CINT(Q, P))**

11. CNCY(ph[I(b)]) = CNCY(ph[E(b)]
   (with proof independent of the previous property)
Proof Steps:
- ph[I(b)] $\subseteq$ ph(b).
- ph[E(b)] $\subseteq$ ph[b].
- Let x $\in$ ph[I(b)].
- Therefore, x $\in$ ph(b) and x $\leftrightarrow$ y for some y $\in$ ph[b].
- Therefore, y $\in$ p[b] and y $\leftrightarrow$ x for some x $\in$ ph(b).
- Therefore, y $\in$ ph[E(b)] and y $\leftrightarrow$ x for some x $\in$ ph[I(b)].
- Therefore, each x of ph[I(b)] $\leftrightarrow$ with some y of ph[E(b)].

- Similarly, we can prove: each x of ph[E(b)] ↔ with some y of ph[I(b)].
- **Therefore, CNCY(ph[I(b)]) = CNCY(ph[E(b)]).**

2.15 Concurrency Hulchuls (C-Hulchuls) of a Body

C-hulchul of a body is the set of concurrency classes of the corresponding p-hulchul of the body. For this purpose, we have the concurrency operator CNCY. Various C-hulchuls of a body b are defined below:

Internal C-hulchul of a Body b
ch(b) = CNCY(ph(b)).

External C-hulchul of Body b
ch[b] = CNCY(ph[b]).

Pure Internal C-Hulchul of a Body b
ch((b)) = CNCY(ph((b))).

Pure External C-Hulchul of a Body b
ch[[b]] = CNCY(ph[[b]]).

Composite C-Hulchul of a Body b
ch[(b)] = CNCY(ph[I(b)]) = CNCY(ph[E(b)]).

Note: We have proved earlier that CNCY(ph[I(b)]) = CNCY(ph[E(b)]) as Property #11 under "properties of CNCY".

Total C-Hulchul of a Body b
ch{b} = CNCY(ph{b})

2.16 Alternative Formulas for Different C-Hulchuls

Now using the properties of CNCY, CDIF, and CINT, we can compute the values of c-hulchuls in alternative ways as shown below. Formulas in the bold are the definitions of c-hulchuls. Formulas not in the bold are alternative expressions for c-hulchuls; it is easier to compute some of the alternative expressions and they also help us gain insight into the c-hulchuls.

Internal C-hulchul of a Body b
ch(b) = CNCY(ph(b)).

External C-hulchul of Body b
ch[b] = CNCY(ph[b]).

Alternative Formulas:

Pure Internal C-Hulchul of a Body b
ch((b)) = CNCY(ph((b)))

$$= CNCY(CDIF(ph(b), ph[b]))$$
$$= CDIF(CNCY(ph(b), CNCY[b]))$$
$$= CDIF(ch(b), ch[b]).$$

Pure External C-Hulchul of a Body b
ch[[b]] = CNCY(ph[[b]]]

$$= CNCY(CDIF(ph[b], ph(b))$$
$$= CDIF(CNCY(ph[b], CNCY(b)))$$
$$= CDIF(ch[b], ch(b)).$$

Composite C-Hulchul of a Body b
ch[(b)] = CNCY(ph[I(b)]) = CNCY(ph[E(b)])

$$= CNCY(CINT(ph(b), ph[b]))$$
$$= CINT(CNCY(ph(b), CNCY(ph[b])))$$
$$= CINT(ch(b), ch[b]).$$

$$= CNCY(CINT(ph[b], ph(b)))$$
$$= CINT(CNCY(ph[b], CNCY(ph(b)))$$
$$= CINT(ch[b], ch(b])).$$

Note: ph[I(b)] = CINT(ph(b), ph[b]), ph[E(b) = CINT(ph[b], ph(b)).

Total Concurrency Hulchul (C-Hulchul) of a Body b
ch{b} = CNCY(ph{b})

$$= CNCY(ph(b) \cup ph[b])$$
$$= CNCY(ph(b)) \cup CNCY(ph[b])$$
$$= ch(b) \cup ch[b]$$

$$= CNCY(ph((b)) \cup ph[I(b)]) \cup CNCY(ph[[b]] \cup ph[E(b)])$$
$$= CNCY(ph((b))) \cup CNCY(ph[I(b)]) \cup CNCY(ph[[b]]) \cup$$
$$CNCY(ph[E(b)])$$
$$= ch((b)) \cup ch[(b)] \cup ch[[b]] \cup ch[(b)]$$
$$= ch((b)) \cup ch[(b)] \cup ch[[b]]$$
Since $CNCY(ph[I(b)]) = CNCY(ph[E(b)])) = ch[(b)]$.

2.17 Examples: Determining C-Hulchuls from P-Hulchuls

As examples, we will derive the values of c-hulchuls from p-hulchuls in more than one way using alternative formulas, where available, for the same c-hulchul.

Suppose
ph(b) = {e11, e21, e31, e42, e53, e63, e74, e84, e85}
ph[b] = {f11, f21, f34, f46, f56, f66, f77, f87}
(Last digit of an event represents the CID of the event.)

ph((b))   = {e42, e53, e63, e85}
ph[[b]]   = {f46, f56, f66, f77, f87}
ph[I(b)] = {e11, e21, e31, e74}
ph[E(b)] = {f11, f21, f34}
ph{b}   = ph(b) ∪ ph[b]
          = {e11, e21, e31, f11, f21, e42, e53, e63, e74, e84, f34, e85, f46, f56, f66, f77, f87}

ch(b)   = CNCY(ph(b))
         = {<1>,<2>,<3>,<4>,<5>}
ch[b]   = CNCY(ph[b])

$= \{<\mathbf{1}>,<\mathbf{4}>,<\mathbf{6}>,<\mathbf{7}>\}$

ch((b)) = CNCY(ph((b)))

$\quad = \{<\mathbf{2}>,<\mathbf{3}>,<\mathbf{5}>\}$

$\quad$ = CDIF(ch(b), ch[b]

$\quad = \{<\mathbf{2}>,<\mathbf{3}>,<\mathbf{5}>\}$

ch[[b]] = CNCY(ph[[b]])

$\quad = \{<\mathbf{6}>,<\mathbf{7}>$

$\quad$ = CDIF(ch[b], ch(b)]

$\quad = \{<\mathbf{6}>,<\mathbf{7}>\}$

ch[(b)] = CNCY(ph[I(b)])

$\quad = \{<\mathbf{1}>,<\mathbf{4}>\}$

$\quad$ = CNCY(ph[E(b)])

$\quad = \{<\mathbf{1}>,<\mathbf{4}>\}$

$\quad$ = CINT(ch(b), ch[b])

$\quad = \{<\mathbf{1}>,<\mathbf{4}>\}$

$\quad$ = CINT(ch[b}, ch(b)]

$\quad = \{<\mathbf{1}>,<\mathbf{4}>\}$

ch{b}   = CNCY(ph{b}) = CNCY(ph(b)) ∪ CNCY(ph[b])

$\quad = \{<\mathbf{1}>,<\mathbf{2}>,<\mathbf{3}>,<\mathbf{4}>,<\mathbf{5}>,<\mathbf{6}>,<\mathbf{7}>\}$

$\quad$ = ch(b) ∪ ch[b]

$\quad = \{<1>,<2>,<3>,<4>,<5>\} ∪ \{<1>,<4>,<6>,<7>\}$

$\quad = \{<\mathbf{1}>,<\mathbf{2}>,<\mathbf{3}>,<\mathbf{4}>,<\mathbf{5}>,<\mathbf{6}>,<\mathbf{7}>\}$

$\quad$ = ch((b)) ∪ ch[(b)] ∪ ch[[b]]

$\quad = \{<2>,<3>,<5>\} ∪ \{<1>\} ∪ \{<4>,<6>,<7>\}$

$\quad = \{<\mathbf{1}>,<\mathbf{2}>,<\mathbf{3}>,<\mathbf{4}>,<\mathbf{5}>,<\mathbf{6}>,<\mathbf{7}>\}$

CHAPTER 3

Prime Hulchul, XP-Hulchul and Concurrency Hulchul

3.01 Introduction

In Chapter 2, we introduced a general theory of concurrency of events with a focus on ticks as events. For this purpose, we also defined some sets of p-ticks as **prime hulchul** by way of examples of events and sets of events. In this chapter, we will discuss prime hulchul and concurrency hulchul in greater details. This chapter, in fact, is in continuation of Chapter 2. We will introduce **extended prime hulchul** of a body b as a set of p-ticks of b or p-ticks of inner and outer bodies of b. We will prove that prime hulchul and extended prime hulchul are equivalent in terms of their concurrency. We will also prove that the total concurrency hulchul (sum of internal concurrency hulchul and external currency hulchul) is the same for all bodies, no matter how small or large a body is. Finally, we will introduce **time outage,** a phenomenon that occurs when a body has only external motion and no internal motion. Time outage is the cause of time dilation. Time outage does not occur continuously, it occurs intermittently, with varying fineness, in case of common bodies.

3.02 Prime Hulchuls (P-Hulchuls)

In Chapter 2, we defined the following eight different types of sets of p-ticks for a body b, called prime hulchuls (or p-hulchuls) of b:

ph(b)   Internal p-hulchul of b

	Set of all independent internal p-ticks of b.
ph[b]	External p-hulchul of b.
	Set of all independent external p-ticks of b.
ph((b))	Pure internal p-hulchul of b.
ph[[b]]	Pure external p-hulchul of b.
ph[I(b)]	Composite internal p-hulchul of b.
ph[E(b)]	Composite external p-hulchul of b.
ph[b, B]	Relative p-hulchul of b in B, where b ⊂ B.
	b ticks in a body b0 where b ⊂ b0 ⊂ B
ph{b}	Total p-hulchul of b.

Definitions of ph((b)), ph[[b]], ph[I(b)], and ph[E(b)] in Terms of Operators CDIF and CINT

$$ph((b)) = CDIF(ph(b), ph[b]) \qquad (3.02.1A)$$
$$ph[[b]] = CDIF(ph[b], ph(b)) \qquad (3.02.1B)$$
$$ph[I(b)] = CINT(ph(b), ph[b]) \qquad (3.02.1C)$$
$$ph[E(b)] = CINT(ph[b], ph(b)) \qquad (3.02.1D)$$

$$ph\{b\} = ph(b) \cup ph[b] \qquad (3.02.1E)$$

3.03 Properties of Prime Hulchuls (P-Hulchuls)

We established the following equations/inequalities related to the above p-hulchuls: (Here, we assume b1 ⊂ b ⊂ B.)

ph(b) and ph[b] are disjoint.

$$ph(b1) \subseteq ph(b) \qquad (3.03.1A)$$
$$ph[b1] \supseteq ph[b] \qquad (3.03.1B)$$
$$ph((b1)) \subseteq ph((b)) \qquad (3.03.1C)$$
$$ph[[b1]] \supseteq ph[[b]] \qquad (3.03.1D)$$

$$ph[b] = ph[b1, b] \cup ph[b] \qquad (3.03.1E)$$
$$ph[b] = ph[b, U] \qquad (3.03.1F)$$
$$ph[b, B] \subseteq ph[b] \qquad (3.03.1G)$$
$$ph[b1, B] \supseteq ph[b, B] \qquad (3.03.1H)$$

$$ph[b1, b] \subseteq ph[b1, B] \qquad\qquad (3.03.1J)$$

3.04 Concurrency of Prime Hulchuls

We defined the CNCY operator to define concurrency hulchuls (c-hulchuls) as follows:

ch(b)	= CNCY(ph(b))	(3.04.1A)
ch[b]	= CNCY(ph[b])	(3.04.1B)
ch((b))	= CNCY(ph((b)))	(3.04.1C)
ch[[b]]	= CNCY(ph[[b]])	(3.04.1D)
ch[(b)]	= CNCY(ph[I(b)])	
	= CNCY(ph[E(b)])	(3.04.1E)
count(ph[b]) = count(ch[b])		(3.04.1F)
count(ph(b)) ≥ count(ch(b))		(3.04.1G)

We proved equivalence of alternative methods to derive values of ch((b)), ch[[b]] and ch[(b)] as follows (definitions from section 3.02):

$$
\begin{aligned}
ch((b)) &= CNCY(ph((b))) \\
&= CNCY(CDIF(ph(b), ph[b])) \\
&= CDIF(CNCY(ph(b)), CNCY(ph[b])) \\
&= CDIF(ch(b), ch[b]) \qquad\qquad (3.04.2A)
\end{aligned}
$$

$$
\begin{aligned}
ch[[b]] &= CNCY(ph[[b]]) \\
&= CNCY(CDIF(ph[b], ph(b))) \\
&= CDIF(CNCY(ph[b]), CNCY(ph(b))) \\
&= CDIF(ch[b], ch(b)) \qquad\qquad (3.04.2B)
\end{aligned}
$$

$$CNCY(ph[I(b)]) = CNCY(ph[E(b)]) \qquad\qquad (3.04.2C)$$

$$
\begin{aligned}
ch[(b)] &= CNCY(ph[I(b)]) \\
&= CNCY(CINT(ph(b), ph[b])) \\
&= CINT(CNCY(ph(b)), CNCY(ph[b])) \\
&= CINT(ch(b), ch[b]) \\
\\
&= CNCY(ph[E(b)])
\end{aligned}
$$

$$= \text{CNCY}(\text{CINT}(\text{ph}[b], \text{ph}(b)))$$
$$= \text{CINT}(\text{CNCY}(\text{ph}[b]), \text{CNCY}(\text{ph}(b)))$$
$$= \text{CINT}(\text{ch}[b], \text{ch}(b)) \qquad\qquad (3.04.2D)$$

3.05 Extended Prime Hulchuls (XP-Hulchuls)

Now we will extend the sets of p-ticks of a body to corresponding sets of extended prime ticks (xp-ticks) of a body. We defined an xp-tick in Chapter 1 as follows: An internal xp-tick of a body b is either an internal p-tick of b or an internal p-tick of an inner body of b. Similarly, an external xp-tick of a body b is either an external p-tick of b or an external p-tick of an outer body of b. Now we define the following xp-hulchuls along with their alternative equivalent definitions:

xph(b): Internal xp-hulchul of a body b.

xph(b) = The set of all independent internal <u>xp-ticks</u> of b. A member of the set will be called an **internal extended prime tick (or xp-tick) of b**. The set xph(b) will be called an **internal extended hulchul (or xp-hulchul) of b**.

Alternative definition of xph(b)

xph(b) = The set of all independent internal <u>p-ticks</u> of all bodies b1 such that b1 $\subseteq$ b.
This is by the definition of external xp-ticks of a body.

Another alternative definition of xph(b)

xph(b) = The set of all independent maximal internal <u>ticks</u> of b.
This is because of Property # 9 of ticks: All independent concurrent internal ticks can be composed into independent maximal internal ticks. All concurrent maximal internal ticks of body form a tree with the outermost body as the root of the tree.

xph[b]: External xp-hulchul of a body b

xph[b] = The set of all independent external <u>xp-ticks</u> of b. A member of the set will be called an **external extended prime tick (or xp-tick) of b**. The set xph[b] will be called an **external extended hulchul (or xp-hulchul) of b**.

Alternative definition of xph[b]

xph[b] = The set of all independent external <u>p-ticks</u> of all bodies B such that b⊆ B.

This is by the definition of external xp-ticks of a body.

Another alternative definition of xph[b]

xph(b) = The set of all independent maximal external <u>ticks</u> of b.

This is because of Property #9 of ticks: All independent concurrent external ticks can be composed into a maximal external tick.

Different Types of Extended Prime Hulchuls (XP-Hulchuls)

** Table 3.05: XP-Hulchul

Set	Description
xph[b]	The set of all independent external p-ticks of all bodies B where b ⊆ B. We will call the set xph[b] the **external extended prime hulchul (xp-hulchul) of the body b** and we will call a member of xph[b] an **external xp-tick of b.**
xph(b)	The set of all independent internal p-ticks of all bodies b1 where b1 ⊆ b. We will call the set xph(b) the **internal extended prime hulchul (xp-hulchul) of the body b** and we will call a member of xph(b) an **internal xp-tick of b.**
xph[b1, b]	The set of all independent external p-ticks of all bodies b0 in b where b1 ⊆ b0 ⊂ b. We will call the set xph[b1, b] the **relative external xp-hulchul of the body b1 in** b.
xph[E(b)]	= CINT(xph[b]), xph(b)) = {x | x ∈ xph[b] and x is concurrent with some y ∈ xph(b)}. We will call the set xph[E(b)] the **composite external xp-hulchul of the body b** and we will call a member of xph[E(b)] a **composite external xp-tick of b.**
xph[I(b)]	= CINT(xph(b)), xph[b]) = {x | x ∈ xph(b) and x is concurrent with some y ∈ xph[b]}. We will call the set xph[I(b)] the **composite internal xp-hulchul of the body** and we will call a member of xph[I(b)] a **composite internal xp-tick of b.**

Set	Description
xph[[b]]	= CDIF(xph[b], xph(b)) = {x | x ∈ xph[b] and x is not concurrent with any member of xph(b)}. We will call the set xph[[b]] the **pure external xp-hulchul of the body b** and we will call a member of xph[[b]] a **pure external xp-tick of** b.
xph((b))	= CDIF(xph(b), xph[b]) = {x | x ∈ xph(b) and x is not concurrent with any member of xph[b] }. We will call the set xph((b)) the **pure internal xp-hulchul of the body b** and we will call a member of xph[[b]] a **pure external xp-tick of** b.
xph{b}	= xph(b) ∪ xph[b] The set of all internal xp-ticks of b and all external xp-ticks of b. We will call the set xph{b} the **total xp-hulchul of b**. **Note**: xph(b) and xph[b] are disjoint in view of the axiom of exclusion of internal and external tick**s**.

The difference between xph[b] and xph[b, B]: xph[b] is a special case of xph[b, B] with b ⊂ B. In case of xph[b, B], b must tick in B whereas in case of xph[b], there is no such restriction on b. In fact, xph[b, B] ⊆ xph[b].

Notes:
1. Suppose x is an internal p-tick of b, x1 is an internal xp-tick of b, and x and x1 are concurrent. Also, suppose y is an external p-tick of b, y1 is an external xp-tick of b, and y and y1 are concurrent. Ticks x and y are concurrent if and only if ticks x1 and y1 are concurrent.
2. CDIF(xph[b], xph(b)) = CDIF(xph[b], ph(b)] since CNCY(xph(b)) = CNCY(ph(b)). Pure external ticks of outer bodies of b are also pure external p-ticks of b since pure external ticks of a body are inherited by all of its inner bodies.
3. CDIF(xph(b), xph[b]) = CDIF(xph(b), ph[b])) since CNCY(xph[b]) = CNCY(ph[b]). Pure internal ticks of inner bodies of b are also pure internal p-ticks of b since pure internal ticks of a body are inherited by all of its outer bodies. Also, see the note after this table.

3.06 Properties of XP-Hulchuls

1. **xph(b) and xph[b] are disjoint** because of Axiom of Exclusion. However, in general, xph(b) and xph[b] are not c-disjoint.

2. **xph[b1] ⊇ xph[b] if b1⊂ b.**
   xph[b] increases or remains the same inner-body-ward.
   Proof Steps:
   - Essentially, it is due to inheritance of external ticks by inner bodies.
   - Let b1⊂ b and x ∈ xph[b].
   - Therefore, x is an external p-tick of some body B such that b ⊆ B.
   - Since b1 ⊂ b ⊆ B, therefore, b1 ⊂ B.
   - Therefore, x ∈ xph[b1].
   - **Therefore, xph[b1] ⊇ xph[b] if b1⊂ b.**

3. **xph(b1) ⊆ xph(b) if b1⊂ b.**
   ph(b1) increases or remains the same outer-body-ward.
   Proof Steps:
   - Essentially, it is due to inheritance of internal ticks by outer bodies.
   - Let b1⊂ b and x ∈ xph(b1).
   - Therefore, x is an internal p-tick of some body b2 such that b2 ⊆ b1.
   - Since b2 ⊆ b1 ⊂ b, therefore, b2 ⊂ b.
   - Therefore, x ∈ xph(b).
   - **Therefore, xph(b1) ⊆ xph(b) if b1⊂ b.**

4. **xph[b, B] ⊆ xph[b] where b ⊂ B.**
   xph[b, B] ⊆ xph[b, B1] where b ⊂ B ⊂ B1.
   xph[b, B] ⊆ xph[b1, B] where b1 ⊂ b ⊂ B.
   Note: The premise here is that if the lineal range [object body, reference body] is larger, then the xp-hulchul is larger, too.
   Proof Steps:
   - We will prove: **xph[b, B] ⊆ xph[b] where b ⊂ B.**
   - Let b ⊂ B and x ∈ xph[b, B].
   - Therefore, x is an external p-tick of b0 in B where b ⊆ b0 ⊂ B.
   - Therefore, x is an external xp-tick of b.
   - **This proves: xph[b] ⊇ xph[b, B] where b ⊂ B.**

- Now we will prove: **xph[b, B] ⊆ xph[b, B1] where b ⊂ B ⊂ B1**.
- Let b ⊂ B ⊂ B1 and x ∈ xph[b, B].
- Therefore, x is an external p-tick of b0 in B where b ⊆ b0 ⊂ B ⊂ B1.
- Since B ⊂ B1, therefore, x is an external p-tick of b0 in B1 where b ⊆ b0 ⊂ B ⊂ B1.
- Therefore, x ∈ xph[b, B1].
- **This proves: xph[b, B] ⊆ xph[B, B1] where b ⊂ B ⊂ B1.**

- Now we will prove: **xph[b, B] ⊆ xph[b1, B] where b1 ⊂ b ⊂ B**.
- Let b1 ⊂ b ⊂ B and x ∈ xph[b, B].
- Therefore, x is an external p-tick of some b0 in B where b1 ⊂ b ⊆ b0 ⊂ B.
- Therefore, x is an external p-tick of b0 in B where b1 ⊂ b ⊆ b0 ⊂ B.
- Therefore, x is an external tick of b1 in B.
- **This proves: xph[b, B] ⊆ xph[b1, B] where b1 ⊂ b ⊂ B.**

5. **xph[b1] = xph[b1, b] ∪ xph[b] if b1⊂ b.**
   Proof Steps:
   - Essentially, it is due to Property #6 of ticks (See Chapter 1), that is, t[b1] = t[b1, b] ∪ t[b] for a single tick.
   - Let b1 ⊂ b.
   - **First, we will prove: xph[b1, b] ∪ xph[b] ⊆ xph[b1]**
   - Let x ∈ xph[b1, b].
   - Therefore, x is an external p-tick of some b0 in b where b1 ⊆ b0 ⊂ b.
   - Therefore, x ∈ xph[b1].
   - Therefore, xph[b1, b] ⊆ xph[b1].
   - Since b1⊂ b, xph[b] ⊆ xph[b1] in view of the property #2 above.
   - **Therefore, xph[b1, b] ∪ xph[b] ⊆ xph[b1].**

   - **Now we will prove: xph[b1] ⊆ xph[b1, b] ∪ xph[b]**
   - Let x ∈ xph[b1].
   - Therefore, x is an external p-tick of some b0 such that b1 ⊆ b0.
   - Since b1 ⊂ b and b1 ⊆ b0, therefore, either b1⊂ b ⊆ b0 or b1 ⊆ b0 ⊂ b.

- Suppose b1⊂ b ⊆ b0. Also, x is an external p-tick of b0. Therefore, x ∈ xph[b].
- Now suppose b1 ⊆ b0 ⊂ b. Also, x is an external p-tick of b0. Therefore, x ∈ xph[b1, b] by definition of xph[b1, b].
- **Therefore, xph[b1] ⊆ xph[b1, b] ∪ xph[b]**
- **This proves: xp[b1] = xph[b1, b] ∪ xph[b].**

6. **xph((b1)) ⊆ xph((b)) if b1⊂ b.**
   xph((b)) increases or remains the same outer-body-ward.
   Proof Steps:
   - Let b1 ⊂ b and x ∈ xph((b1)).
   - Therefore, x ∈ xph(b1) and x is not concurrent with any member of xph[b1].
   - Therefore, x ∈ xph(b) and x is not concurrent with any member of xph[b], in view of properties #1 and 2 above.
   - Therefore, x ∈ xph((b)).
   - **This proves: xph((b1)) ⊆ xph((b)).**

7. **xph[[b]] ⊆ xph[[b1]] if b1⊂ b.**
   xph[b] increases or remains the same inner-body-ward.
   Proof Steps:
   - Let b1⊂ b and x ∈ xph[[b]].
   - Therefore, x ∈ xph[b] and x and x is not concurrent with any member of xph(b).
   - Therefore, x ∈ xph[b1] and x is not concurrent with any member of xph(b1), in view of properties #1 and 2 above.
   - Therefore, x ∈ xph[[b1]].
   - **This proves: xph[[b]] ⊆ xph[[b1]].**

8. **xph{b1} ⊆ xph{b} if b1⊂ b.**
   Proof Steps:
   - Let b1⊂ b.
   - Therefore, xph[b1] = xph[b1, b] ∪ xph[b] in view of Property #4 above.
   - Or xph[b1] ⊆ xph(b) ∪ xph[b], since xph[b1, b] ⊆ xph(b).
   - Or xph(b1) ∪ xph[b1] ⊆ xph(b1) ∪ xph(b) ∪ xph[b].
   - Or xph(b1) ∪ xph[b1] ⊆ xph(b) ∪ xph[b], since xph(b1) ⊆ xph(b).
   - **This proves: xph{b1} ⊆ xph{b}.**

The above property is a strong statement since internal xp-hulchul xph[b] increases or remains the same inner-body-ward and external xp-hulchul xph(b) increases or remains the same outer-body-ward, whereas total xp-hulchul increases or the remains same outer-body-ward in line with internal xp-hulchul. Why is there this preference for internal xp-hulchul over external xp-hulchul? This is due to the fact that internal ticks have a tree structure, whereas external ticks have a lineal structure. (See the Figure 3.06A and Figure 3.06B comparing structures of internal and external ticks.) As a result the number of internal ticks dominates the number of external ticks; that is, from a body b1 to b, where b1⊂ b, the increase in internal hulchul is more than the decrease in external hulchul.

In the next section, we will establish another measure of hulchul, to be called concurrency, and prove that concurrency of xph{b1} and xph{b} are equal, in fact, equal for any two bodies.

** **Figure 3.06A**

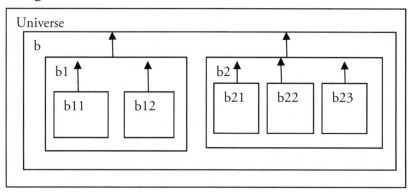

Internal ticks of a body b have a tree structure.

** **Figure 3.06B**

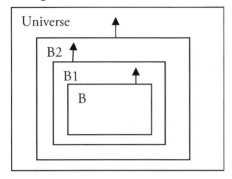

External ticks of a body B have a lineal structure.

9. **xph[b] = xph[b, U]**
    It is based on t[b] = t[b, U]).

10. **xph[U] is null**
    Since U has no external ticks.

11. **xph(e) is null where e is an elementary body**
    Since an elementary body e has no internal tick.

The Implication of the Range of Lineal Bodies of a Tick for xp-Hulchul

In Chapter 1, we defined the range of two lineal bodies for a tick as the pair of object and reference bodies. Now we interpret this for xp-hulchuls. We note the following (we assume: b2 ⊂ b1 ⊂ b ⊂B):

Rules for Change in Relative Hulchul

1. If the reference body is the same, then relative xp-hulchul increases with respect to the object body inner-body-ward.

2. If the object body is the same, then relative xp-hulchul increases with respect to the reference body outer-body-ward.

3. That is, a wider range of a pair of object body b and reference body B implies larger xp-hulchul xph[b, B]. This is due to inheritance of external ticks by inner body and internal tick by outer body.

4. A range of a pair of object body b̂ and reference body B, [b, B] behaves similarly as does an interval of real numbers (x, y) or [x, y] where x and y are real numbers and x < y.

5. xph[b] is the same as xph[b, U].

Examples

- xph[b2] $\supseteq$ xph[b2, b] $\supseteq$ xph[b1, b]
  This is due to the property of inheritance of external ticks by inner body. Range: [b2, U] $\supset$ [b2, b] $\supset$ [b1, b]

- xph[b1, b] $\subseteq$ xph[b1, B] $\subseteq$ xph[b1]
  This is due to the property of inheritance of internal ticks by outer body. Range: [b1, b] $\subset$ [b1, B] $\subset$ [b1, U]

- xph[b2] $\supseteq$ xph[b2, B] $\supseteq$ xph[b2, b] $\supseteq$ xph[b1, b]
  xph[b2] $\supseteq$ xph[b2, B] $\supseteq$ xph[b1, B] $\supseteq$ xph[b, B]

  Where b2 $\subset$ b1 $\subset$ b $\subset$ B
  Range: [b2, U] $\supset$ [b2, B] $\supset$ [b2, b] $\supset$ [b1, b]

3.07 Concurrency Ticks (C-Ticks) and Concurrency Hulchul (C-Hulchul)

We will use the operators CNCY, CDIF and CINT to determine corresponding c-hulchuls from xp-hulchuls. First, we note:

$$CNCY(xph(b)) = CNCY(ph(b)) \qquad (3.07.1A)$$
$$CNCY(xph[b]) = CNCY(ph[b]) \qquad (3.07.1B)$$

Equation (3.5.1A) is true, due to Property #4 of ticks (xp-ticks): an internal xp-tick t of a body b is either an internal p-tick of b or it is concurrent with an internal p-tick of b.

Similarly, Equation (3.5.1B) is true, due to the property of external xp-ticks that an external xp-tick t of a body b is either an external p-tick of b or it is concurrent with an external p-tick of b.

Therefore,

ch(b) = CNCY(xph(b)) = CNCY(ph(b)) (3.07.2A)
ch[b] = CNCY(xph[b]) = CNCY(ph[b]) (3.07.2B)

ch(b) and ch[b] can also be defined from the first principle without using the concurrency operator CNCY, as follows:

ch(b): Divide all independent internal p-ticks in xph(b) into classes of concurrent p-ticks. This implies that all p-ticks in a class are concurrent but no two p-ticks, chosen one each from the two different classes, are concurrent.

ch[b]: Divide all independent external p-ticks in xph[b] into classes of concurrent p-ticks. This implies that all p-ticks in a class are concurrent but no two p-ticks, chosen one each from the two different classes, are concurrent.

To compute c-hulchuls from pure internal xp-hulchul, pure external xp-hulchul and composite xp-hulchul, we will use the following procedure established and proven in Chapter 2:

xph((b)) = CDIF(xph(b), xph[b]))
xph[[b]] = CDIF(xph[b], xph(b)))
xph[I(b)] = CINT(xph(b), xph[b])
xph[E(b)] = CINT(xph[b], xph(b))

ch((b)) = CNCY(xph((b)))
= CNCY(CDIF(xph(b), xph[b])
= CDIF(CNCY(xph(b)), CNCY(xph[b]))
= CDIF(ch(b), ch[b])   (3.07.3A)

Since ph[b] and xph[b] are c-equivalent and ph(b) and xph[b] are c-equivalent, therefore,

CDIF(xph[b], xph(b)) = CDIF(xph[b], ph(b))

CDIF(xph(b), xph[b]) = CDIF(xph(b), ph[b])

$$ch[[b]] = CNCY(xph[[b]])$$
$$= CNCY(CDIF(xph[b]), xph(b))$$
$$= CDIF(CNCY(xph[b]), CNCY(xph(b)))$$
$$= CDIF(ch[b], ch(b)) \qquad (3.07.3B)$$

$$CNCY(xph[I(b)]) = CNCY(xph[E(b)] \qquad (3.07.3C)$$

This is because each member of xph[I(b)] is concurrent with some member of xph[E(b)] and vice versa as proven in Chapter 2.

$$ch[(b)] = CNCY(xph[I(b)])$$
$$= CNCY(CINT(xph(b), xph[b]))$$
$$= CINT(CNCY(xph(b)), CNCY(xph[b]))$$
$$= CINT(ch(b), ch[b])$$

$$= CNCY(xph[E(b)])$$
$$= CNCY(CINT(xph[b], xph(b)))$$
$$= CINT(CNCY(xph[b]), CNCY(xph(b)))$$
$$= CINT(ch[b], ch(b)) \qquad (3.07.3D)$$

$$CNCY(xph[b1, b]) = CNCY(ph[b1, b])) \qquad (3.07.3E)$$

This is because: ph[b1, b] ⊆ xph[b1, b]. On the other hand, if x belongs to xph[b1, b], then either x belongs to ph[b1, b] or x is concurrent with some tick from ph[b1, b].

We have already proven the following equation as Property #4 of xp-hulchul.

$$xph[b1] = xph[b1, b] \cup xph[b] \qquad (3.07.3F)$$

This means c-hulchul of a p-hulchul and corresponding xp-hulchul of a given type are the same.

We should also note:

$$ch(b) = ch((b)) \ U \ ch[(b)]$$
$$ch[b] = ch[[b]] \ U \ ch[(b)]$$
$$ch\{b\} = ch(b) \ U \ ch[b] = ch((b)) \ U \ ch[(b)] \ U \ ch[[b]]$$

ch(b), ch[b], ch[b1, b], ch((b)), ch[[b]] and ch[(b)] and ch{b} all are, in fact, subsequences of ch(U), and therefore in the same order as that of the c-ticks in ch(U).

** **Figure 3.07**

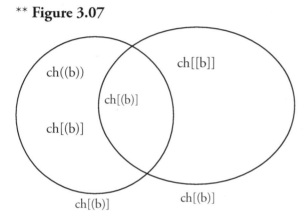

Concurrency Ticks (C-ticks)

A member of any c-hulchul, in general, will be called a **concurrency tick** or **c-tick**. We will also call a c-tick as **concurrency instant** or **c-instant**. Specific names for a c-tick in some of the c-hulchuls are:

External c-tick: Member of external c-hulchul ch[b].
Internal c-tick: Member of internal c-hulchul ch(b).
Composite c-tick: Member of composite c-hulchul ch[(b)].
Pure external c-tick: Member of external c-hulchul ch[[b]].
Pure internal c-tick: Member of internal c-hulchul ch((b)).

Count of members in a Set: We will denote the number of members in a set S by **count(S)**. As a special case, we will also denote the counts of the sets of all prime hulchuls and concurrency hulchul by replacing "ph" by "PH" for the count of corresponding prime hulchul, "xph" by

"XPH" for the count of corresponding xp-hulchul, and by replacing "ch" by "CH" for the count of corresponding concurrency hulchul. For example, we will denote count of ph(b) by PH(b), count of xph(b) by XPH(b) and count of ch(b) by CH(b).

3.08 Universal C-Ticks and Universal C-hulchul

The set xph(U) and ch(U) for the body U are of special interest.

xph(U)
= Set of all independent internal p-ticks of all bodies in the U body.
= Set of all independent external p-ticks of all bodies in the U body.

This is because an internal p-tick of U is also an external p-tick of some inner body of U. An internal xp-tick of U is an internal p-tick of some inner body of U.

ch(U) = CNCY(xph(U))

We will call the set ch(U) the **universal c-hulchul** and a member of ch(U) a **universal c-tick.** ch(U) is the sequence of all c-ticks in the universe.

Question: How large is the set xph(U)? What is the number of all independent p-ticks or c-ticks in the U body, say, per microsecond?

We do not know but these should be astronomical numbers.

Question: Are there any events in the universe that do not involve some kind of motion?

If the answer to the above question is "NO", then we ask the following question:

Question: Is there any motion without involving some p-tick?
No.

Question: Are universal c-ticks precisely the instants of time? Equivalently, we may ask: Does there exist at least one p-tick in xph(U) corresponding to each instant of time?

Presumably, occurrence of an event always involves an instant of time.

We do not know yet the definite answers to all of the above questions. We do not know yet whether universal c-ticks can be used as instants of physical time but they appear to have the potential for this role. However, we will use universal c-ticks as **instants of time** *at least in a mathematical sense.*

Now we will define a time interval of time in terms of universal c-ticks.

3.08.1 Hulchul Domain (H-domain)

Usually, we use time as either as an instant of time or as a time-interval between two instants of time. However, a time-interval is not good enough for the purpose of this research to deal with, for examples, intermittent inner-outer body relationship between bodies, intermittent motion and intermittent time. A time-interval is essentially a continuous subsequence of the instants of time. For intermittent relationship we need discontinuous subsequence of instants of time. For this purpose, we will introduce a term to be called **hulchul domain** (or *h-domain)* as defined below.

Continuous and Discontinuous h-domain: We will call a subsequence of ch(U) a **hulchul domain** (or **h-domain**). We will call an h-domain a **continuous h-domain** or **discontinuous h-domain** according as the underlying subsequence of ch(U) is continuous or discontinuous.

Examples: All c-hulchuls of a body b are h-domains of b. All c-hulchuls of a common body b, except total c-hulchul ch{b}, are discontinuous h-domains; total c-hulchul ch{b} may be a continuous or a discontinuous h-domain.

Note: A sequence is an ordered set. If S1 is a subsequence of a sequence S, then S1 is a subset of S, and it retains the order of S. We define a

continuous subsequence S1 of S as follows: If x and y both belong to S1, then all members of S between x and y also belong to S1. If the subsequence is not continuous, then it is said to be a **discontinuous subsequence**.

Density of h-domain: Suppose h1 is an h-domain and S1 is the corresponding subsequence of ch(U). Let x and y be the lowest (oldest) c-tick and the highest (newest) c-tick in h1. Then the percentage of the number of c-ticks in S1 to the number of c-ticks in ch(U) between x and y will be called the **density of the h-domain** h1. The density of a continuous h-domain is 100 %.

*Now onward, all statements on hulchuls involving more than one instant of time, such as p-ticks, p-hulchul, xp-ticks, xp-hulchul, c-ticks and c-hulchul, will assume an underlying h-domain unless mentioned otherwise. However, in general, an h-domain should be **sufficiently large**. A pattern in the distribution of c-ticks of different types of a common body does not emerge unless the h-domain is sufficiently large.*

"Sufficiently large" h-domain is a relative term here. In terms of number of c-ticks, a million is sufficiently large. In terms of time, even a microsecond is more than sufficiently large.

A simpler notation for counts of different c-hulchuls

At times, we may find it handy to use following simpler notations for the counts of the following c-hulchuls when the underlying body b is known:

I = count(ch((b)))	Number of pure internal c-ticks of b
C = count(ch[(b)])	Number of composite c-ticks of b
E = count(ch[[b]])	Number of pure external c-ticks of b
Ic = I + C	
Ic = count(ch(b))	Number of internal c-ticks of b
Ec = E + C	
Ec = count(ch[b])	Number of external c-ticks of b
ICE = I + C + E	
$ICE2$ = Ic + Ec	

The difference between ICE and ICE2: ICE is the number of all distinct c-ticks in ch{b} whereas ICE2 is the sum of the number of all distinct c-ticks in ch(b) plus the number of all distinct c-ticks in ch[b]. We will use ICE2 in several computations.

3.09 Properties of Concurrency Hulchuls (C-Hulchuls)

We assume b1⊂ b.

1. **Inheritance Properties of C-hulchuls:**
   It is based on: CNCY(S1) ⊆ CNCY(S) if S1 ⊆ S where S1 and S are sets of events.
   - **ch(b1) ⊆ ch(b)**; internal c-hulchul of b, ch(b), increases or remains the same outer-body-ward.
   - **ch((b1)) ⊆ ch((b))**; pure internal c-hulchul of b, ch((b)), increases or remains the same outer-body-ward.
   - **ch[b1] ⊇ ch[b]**; external c-hulchul of b, ch[b], decreases or remains the same outer-body-ward.
   - **ch[[b1]] ⊇ ch[[b]]**; pure external c-hulchul of b, ch[[b]], decreases or remains the same outer-body-ward.
   - **ch{b1} ⊆ ch{b}**; total c-hulchul; this is because: xph{b1} ⊆ xph{b}. (We will propose a stronger statement than this as **Proposition #1** a little later.)
   - **Composite** c-hulchul does not have any of the above properties.

2. **Properties of the Counts of C-hulchuls:**
   - **count(ch(b1)) ≤ count(ch(b))**; count(ch(b)) increases or remains the same outer-body-ward.
   - **count(ch((b1))) ≤ count(ch((b)))**; count(ch((b))) increases or remains the same outer-body-ward.
   - **count(ch[b1]) ≥ count(ch[b])**; count(ch[b]) decreases or remains the same outer-body-ward.
   - **count(ch[[b1]]) ≥ count(ch([[b]]))**; count(ch[[b]]) decreases or remains the same outer-body-ward.
   - **count(ch[b]) ≤ count(xph[b])**
   - **count(ch[b]) = count of maximal external ticks of b**

- **count(ch[b]) = count(ph[b])**
- **count(ch[[b]]) = count(ph[[b]])**
- **count(ch[[b]]) = count(xph[[b]])**
  (See 3.09.1 below.)
- **count(ch(b)) ≤ count(xph(b))**
- **count(ch((b))) ≤ count(xph((b)))**

3.09.1 Pure and composite xp-ticks

Are pure and composite xp-ticks of a body b the same as the corresponding p-ticks of b or are they different?

Concurrent external xp-ticks of a body b form a maximal external tick. All outer bodies of b, as object bodies, in the maximal tick of b have composite ticks and none of them has a pure external tick. Therefore, pure external xp-ticks of a body are the same as the pure external p-ticks of b.

Similar rule applies to internal xp-ticks.

Pure and composite c-hulchuls of a body b are computed directly from ch(b) and ch[b] by using the alternative methods as discussed in Section 3.7 (see equations: 3.7.3A, 3.7.3B and 3.7.3D).

3.10 Motion and Time in Terms of C-Ticks

Motion of a Body in Terms of External Ticks
Generally, motion is associated with distance traveled. **We believe motion can also be defined in terms of external ticks of a body.** As time has instants of time, motion can also be associated with instants of motion. Which external ticks, out of several types, will we associate with motion? Let us first recap important facts about external p-ticks, xp-ticks and c-ticks of a body b:

- We have two sets of external ticks: ph[b] and xph[b] defined in different ways. Their members are called p-ticks and xp-ticks respectively.

- ph[b] = The set of all independent external p-ticks (p-ticks) of b.
- xph[b] = The set of all independent external extended prime ticks (xp-ticks) of b where an external xp-tick of b is either an external p-tick of b or an external p-tick of an outer body of b.
- xph[b] = The set of all maximal external ticks of b.
- **All concurrent external ticks of body b form a unique maximal external tick of b.**

- We have only one set of external c-ticks: ch[b].
- ch[b] = CNCY(ph[b]) = CNCY(xph[b]).
- count(ph[b]) = count(ch[b]).
  There is a one-to-one correspondence between ph[b] and ch[b].
- count(xph[b]) ≥ count(ch[b])
  There is a many-to-one correspondence between xph[b] and ch[b].
- count(maximal external ticks of a body) = count(ch(b)).
  There is a one-one correspondence between maximal external ticks of b and ch[b].
- During a single universal c-tick, a maximal external tick of b involving more than one independent p-tick exerts more than one **kinetic pull** on b than a single p-tick does.

We will assume that an external c-tick represents the external motion of the body b. A c-tick is equivalent to a maximal external tick of b since there is a one-to-one correspondence between maximal external ticks and c-ticks of b. We will call each external c-tick of b an **instant of motion** *of b.*

Notwithstanding the above statement, it is possible that the number of independent external p-ticks of a body may have their own physical significance; we don't know at this stage of this research what it is.

Body-Specific Time (BS-Time) in Terms of Internal C-Hulchul

We will assume ch(b) as the **body-specific time (or bs-time)** for the body b. We will call the bs-time for the U body as **universal time** or **u-time**.

Density of Time: ch(b) is a subsequence of ch(U) and therefore it is an h-domain. ch(b) as an h-domain is continuous only in case of the U body; in case of a common body, ch(b) is a discontinuous h-domain. ch(b) as an h-domain has density; we will call the density of ch(b) as the **density of body-specific time,** or **density of bs-time** or simply **density of time**. The density of the u-time is 100 % and only u-time can have 100 % density of time.

An example of concurrency of xp-ticks

** Figure 3.10

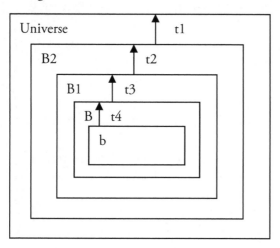

In Figure 3.10 there are four external p-ticks of b. PH[b] = 1 and XPH[b] = 4. Any number of them may be concurrent. For example, if no two of them are concurrent, then CH[b] = 4; if all are concurrent, then CH[b] = 1.

3.11 Proposition #1

In a given h-domain, total c-hulchul of any two bodies is the same, no matter how small or large either body is. That is, for any two bodies b and q, lineal or non-lineal:

$$ch\{b\} = ch\{q\} = ch\{U\} = ch(U) \quad (3.11)$$

We have been able to prove

$$ch\{b1\} \subseteq ch\{b\} \text{ where } b1 \subset b \text{ (3.11A)}$$

by using

$$xph\{b1\} \subseteq xph\{b\}$$
$$CNCY(xph\{b1\}) \subseteq CNCY(xph\{b\})$$

But we are unable to prove the following logically:

$$ch\{b1\} \supseteq ch\{b\} \text{ where } b1 \supseteq b \text{ (3.11B)}$$

It turns out tougher to prove (3.11B). Maybe, we need to be aware of certain physical characteristics related to hulchul. However, we can prove (3.11) by invoking the following two axioms related to physical properties of a body:

3.11.1 The Two Axioms of Elementary Bodies

Axiom 1 of Elementary Bodies: Each body is either an elementary body or it has at least one elementary body as an inner body during each universal c-tick.

The first axiom of elementary bodies implies that given any common body b and a universal c-tick x, there exists at least one elementary body e such that $e \subset b$; e depends on b and x. For the next universal c-tick y, e may not be an inner body of b or e may not even exist—it might have decayed. In case, e is no longer an elementary body of b, then some other elementary body must be an inner body of b.

Axiom 2 of Elementary Bodies: Each elementary body has a unique external p-tick corresponding to each universal c-tick.

Here, the required external p-tick of the underlying elementary body e need not be a native tick; it may be inherited from one of the outer bodies, say b, of e. That is, t[e] = t[e, b] ∪ t[b] for a given universal c-tick x, where at least one of t[e, b] and t[b] is non-empty. This implies:

$$ch\{e\} = ch\{U\} \qquad \text{(3.11C)}$$

Intuitive justification for the second axiom of elementary is as follows: Outer-body-ward, the U body is the maximum body and inner-body-ward an elementary body is a minimal body. Outer-body-ward, the maximum body U has maximum internal c-hulchul. Similarly, inner-body-ward, it is reasonable to assume that an elementary body, being an extreme body, has maximum external c-hulchul.

Neutrinos may have the most native motion or native ticks: Neutrinos have both native and inherited motion. However, among material elementary particles, neutrinos appear to have the most native motion or native ticks.

Proof of Proposition #1:
For any two bodies b and q:

$$ch\{b\} = ch\{q\} = ch\{U\} = ch(U)$$

Proof Steps:

- We already know for any elementary body e, $ch\{e\} = ch\{U\}$.
- We will now prove $ch\{b\} = ch\{U\}$ for any common body b.
- Let x be a universal c-tick.
- The first axiom of elementary bodies implies that there exists an elementary body e such that $e \subset b$, where e depends on body b and universal c-tick x.
- The second axiom of elementary body implies that e has a unique external p-tick during universal c-tick x. That is: $t[e] = t[e, b] \cup t[b]$ where at least one of $t[e, b]$ and $t[b]$ is a non-empty tick using Property #6 of ticks.
- This means b has either an internal tick because of $t[e, b]$ or b has an external tick because of $t[b]$ or both.
- This means b has a p-tick (internal, external or composite) during x.
- This means b has a c-tick (internal, external or composite) during any universal c-tick x.
- That is, $ch\{U\} \subseteq ch\{b\}$.
- Also, $ch\{b\} \subseteq ch\{U\}$ since $b \subset U$.
- Therefore, $ch\{b\} = ch\{U\}$.

- Similarly, we can prove ch{q} = ch{U}.
- **Therefore, ch{b} = ch{q} = ch{U} = ch(U) for any two bodies b and q.**

3.11.2 Axiom 2 of Elementary Bodies

Axiom 2 of elementary bodies is both a necessary and a sufficient condition for the Proposition #2 to be true.

Suppose Proposition #2 is true and e is an elementary body. Then

$$ch\{e\} = ch(U)$$

This means given any universal c-tick x, e has a unique external p-tick tick during x. This is precisely Axiom 2 of elementary bodies.

3.12 Implications of Proposition #1

1. **For a given h-domain, the value of count(ch{b}), the total number of c-ticks for any body b, is a universal constant**; it does not depend on the body b but it does depend on the h-domain.

2. For a photon or light p, ch{p} = ch[p] since p has no external c-hulchul. Therefore, count(ch[p]) is a constant. Therefore, **speed of light or photon is constant.** We should remember that all external ticks of a photon are native, whereas it is not so in the case of material elementary particles.

3. Since ch{p} = ch{U}, therefore, ch{p} contains all universal c-ticks. Therefore, a photon does not pause. Therefore, a photon must change direction rapidly, in fact, after each c-tick. **So the motion of a photon is not straight; rather, it is wavy straight.** (This was the intuition that led us to define motion as a sequence of occurrences of the two steps: Move and pause/change direction.)

4. Since count(ch{b}) = count(ch((b)) + count(ch[(b)]) + count(ch[(b)]) and count(ch{b}) = count(ch(U)). Therefore,

$$\mathbf{I + C + E = CH(U)}$$

5. **Any p-tick of any body must be concurrent with some p-tick of any other body.** The two corresponding p-ticks may not be of the same type—one may be an internal p-tick of one body and the other may be an external p-tick of the other body.

Note: Proposition #1 should not be a complete surprise since internal c-hulchul ch(b) increases and external c-hulchul ch[b] decreases outer-body-ward and the total c-hulchul = ch(b) ∪ ch[b].

The Crucial Difference between Internal and External C-hulchuls

Though
$$CNCY(ph(b)) = CNCY(xph(b))$$
$$CNCY(ph[b]) = CNCY(xph[b])$$

But
$$count(CNCY(ph(b))) \leq count(xph(b))$$
Or $count(ch(b)) \leq count(xph(b))$

$$count(CNCY(ph[b])) = count(ph[b])$$
Or $count(ch[b]) = count(ph(b))$

Due to the axiom of independent concurrent external ticks of a body b, an external tick of b involves concurrent external p-ticks only of lineal bodies but an internal tick may involve concurrent internal p-ticks of non-lineal bodies as shown in the Figure 3.06A and Figure 3.06B.

3.13 Dissecting Total C-Hulchul of a Body

1. **The same c-hulchul for all bodies**: Probabilistically, total c-hulchul of all bodies is the same in all large h-domains but the types of c-ticks in the c-hulchul may change from one body to another.

2. **Order of c-ticks**: The order of c-ticks in the c-hulchul remains the same for all bodies. (See the note in Section 2.02 regarding order of events as observed by two different observers. We assume order of events is independent of observers.)

3. **Composite c-ticks**: A composite c-tick of a body b in the total c-hulchul has two parts: an external c-tick of b and an internal c-tick of b, the two c-ticks must be consecutive and concurrent. The internal c-hulchul and the external c-hulchul of b overlap due to the presence of composite c-ticks. A composite c-tick is one unique c-tick in the total c-hulchul but two different c-ticks in the context of internal c-hulchul and external c-hulchul.

4. **Change in types of c-ticks**: C-ticks of different types are interspersed into the total c-hulchul randomly subject to certain rules. Types of c-ticks may change in some situations and may remain the same in some other situations. There are five types of c-ticks: external, internal, pure external, pure internal, and composite.

    *   **External c-tick type**: An external c-tick of a body b must remain an external c-tick for an inner body of b but it may change to an internal c-tick of an outer body of b.

        Pure external c-tick type: A pure external c-tick of a body b must remain a pure external c-tick for an inner body of b but a composite c-tick of b may also change to pure external c-tick for an inner body of b.

    *   **Internal c-tick type**: There are rules, similar to the above two rules, for internal c-ticks.

    *   **Composite c-tick type**: In general, there is no rule for the composite type of c-ticks since a composite c-tick is formed of an internal c-tick and external c-tick and the rules for external c-ticks work inner-body-ward and the rules for internal c-ticks work outer-body-ward.

- **Comparison of c-hulchuls of two non-lineal bodies**: If we need to compare total c-hulchuls of two non-lineal bodies p and q, then we must first identify some common outer body b of p and q (U is always one such common outer body). Then compare p with b, and then compare q with b.

- **c-ticks, maximal ticks and p-ticks**: Suppose b is a body. A c-hulchul is a c-set but a p-hulchul is not a c-set in general. Count(ph{b}) ≥ count(ch{b}). There is one-one correspondence between external c-ticks of b and maximal external ticks of b and between external c-ticks of b and p-ticks of b. There is one-one correspondence between internal c-ticks of b and maximal trees of internal p-ticks of b.

Convergence of Three Concepts: Internal C-hulchul of the U body, Universal Time and Motion of Light: It appears that the three concepts converge into one and the same concept. Further, if ch(U) = uTu, where u is the number of universal c-ticks per unit of time, and Tu is the number of units of time, then u and c, the speed of light, are the same except they may be using different units of time.

3.14 Time Outage and Time Dilation

We will assume here that the universal c-ticks are precisely the instants of time. Consider the following 50 consecutive universal c-ticks for the body U and corresponding 50 c-ticks for a common body b. Each X, Y and Z indicates occurrence of a pure internal c-tick, a pure external c-tick and a composite c-tick respectively. The body U has all X's only as it can have only pure internal c-ticks.

```
             1         2         3         4         5
    12345678901234567890123456789012345678901234567890
U   XXXXXXXXXXXXXXXXXXXXXXXXXXXXXXXXXXXXXXXXXXXXXXXXXXXX
b   YXXXXYZXYXYZYYYYXXYYZYYYYXZXXXYZXXXXXZZYYXXXXYYXXX
```

Comparison of c-ticks of a common body b with 50 consecutive universal c-ticks:

We have the following counts of c-ticks in different categories for the body b:

ICE = 50	Total number of c-ticks
I = 23	Number of pure internal c-ticks
E = 20	Number of pure external c-ticks
C = 7	Number of composite c-ticks
Ic = 30	Number of internal c-ticks
Ec = 27	Number of external c-ticks

Note: This is not empirical data; this is just to demonstrate the concept.

The body b experiences the above number of c-ticks in the three categories with a total of 50 c-ticks. The body U experiences only pure internal c-ticks as it always does. Our interpretation of the various numbers of c-ticks in the three categories is as follows:

1. The body b does not experience the instants of motion during the 23 pure internal c-ticks; it does experience instants of motion during each of the other 27 c-ticks.

2. Similarly, the body b does not experience instants of time during the 20 pure external c-ticks; we will call this phenomenon **time outage;** it does experience the instants of time during each of the other 30 c-ticks. **Thus the time outage of a body b is ch[[b]], the pure external c-hulchul of b.** We will discuss it more in section 3.14.1.

3. Since a common body has pure external c-hulchul intermittently, therefore its time outage is also intermittent with varying fineness of intermittency.

4. Here, the motion of the body b is not continuous; it is discrete and intermittent since external c-ticks are interspersed with pure

internal c-ticks. A similar interpretation is for the body-specific time for b.

5. The body b experiences instants of both motion and time during the 7 composite c-ticks.

6. Each body, except the U body, appears to have pure motion and therefore, it has time outage even though it may be in a state of relative rest.

State of motion can be both absolute and relative; state of rest can be relative only. Bodies seemingly even in a state of relative rest may have time outage (and time dilation).

We believe that **time outage** and **time dilation** are either the same phenomenon or time outage is the cause of time dilation. For most of the common bodies, time outage is not continuous; it is intermittent. Time outage and time dilation are based on different premises as outlined below:

** **Table 3.14**: Time Outage versus Time Dilation

Time Outage	Time Dilation
Motion is absolute or relative.	Motion is relative.
Based on absolute motion only (at this stage of the research).	Based on relative motion.
Motion is discrete and intermittent.	Motion is continuous.
Time is discrete and intermittent.	Time is used as if it is continuous.
Common bodies, other than the body U, have **some** amount of time outage in all sufficiently large h-domains.	A body has time dilation only when it has relative motion.
Motion and time are two views of the same physical concept.	Motion and time are two different physical concepts.

Note: Twins Paradox does not occur in case of Time Outage as Time Outage is based on absolute motion. Even in a case of hulchulic relative motion, we know exactly the body that is moving and the one that may not be moving. For example in the case ch[b, B], b moves. B may also move, in that case, motion of b includes (inherits) the motion of B.

3.14.1 Time Outage

Time outage of a body b is a partial or full absence of the body-specific time (bs-time) of b; it may be an intermittent and not a continuous absence. Internal c-hulchul of b is the bs-time of b. Therefore, the full absence of bs-time of b is: ch{b} – ch(b), which is the same as ch[[b]], since ch{b} = ch(b) ∪ ch[[b]]. This means the full **time outage** of a body b is ch[[b]], the pure external c-hulchul of b. A partial time outage of b is a set of pure external c-ticks of b. Time outage appears to be the cause of time dilation but the two are different conceptually.

We will now formally define time outage.

Time Outage C-Tick of a Body: An external c-tick x of a body b will be called a **time outage c-tick** of b, if b does not experience an internal c-tick of b during x. That is x is a pure external c-tick. That is, a time outage c-tick of b experiences the universal time (u-time) and external c-tick (motion) but it does not experience the body-specific time.

Time Outage of a Body: The set of time outage c-ticks of b during a given h-domain will be called the **time outage** of b. So the time outage is ch[[b]].

Note: Time outage is absolute in the case of an absolute external c-hulchul ch[b] (absolute motion), or it is relative in the case of a relative c-hulchul ch[b, B] (relative motion).

3.15 Special Cases of C-Hulchuls

3.15.1 Uniform C-Hulchul H-domain-wise

Definition: We will say a body b has uniform c-hulchul h-domain-wise if the ratio of c-hulchul value of each of I, C and E to the total c-hulchul

ICE is the same for all large h-domains. For example, value of I/ICE remains the same h-domain-wise. The same is true for E/ICE and C/ICE. But the values of I/ICE, E/ICE and C/ICE may be different from each other.

Obviously, in case of uniform c-hulchul, each of Ic and Ec will also have the same ratio to the total c-hulchul h-domain-wise as do I, C and E have. However, there is no uniformity in c-hulchul body-wise. Even h-domain-wise, a body may not have uniform c-hulchul, for example, in case of accelerated motion. In general, there does not seem to be uniformity in c-hulchul body-wise (from body to body) for the same h-domain. For example: I/ICE is not the same from one body to another.

Unless mentioned otherwise, we will assume h-domain-wise uniform c-hulchul.

3.15.2 Body-wise Change in C-hulchuls

It appears that change in c-hulchuls is not uniform body-wise, it may be exponential. How exactly c-hulchul changes outer-body-ward or inner-body-ward, we did not yet determine. We will discuss it more in Chapter 4.

3.15.3 Middle Bodies

Since internal c-hulchul Ic increases and external c-hulchul Ec decreases outer-body-ward, there must exist bodies m having Ic = Ec. Such bodies may be real or hypothetical; we will call them **middle bodies.** We will call the common external/internal c-hulchul of a middle body a **middle c-hulchul**.

Let $b1 \subset m \subset b$.
A middle body m implies:

$$CH(m) = CH[m]$$
$$CH((m)) = CH[[m]] \text{ since Ic + E = Ec + I = ICE and Ic = Ec}$$

CH(b1) ≤ CH[b1] if b1 ⊂ m (3.15.3A)
CH(b) ≥ CH[b] if b ⊃ m (3.15.3B)

Also, because of inheritance properties
ch(b1) ⊆ ch(b)
ch[b1] ⊇ ch[b]
CH[b1] ≤ CH[b]
CH[b1] ≥ ch[b]

We may note from the inequalities (3.15.3A) and (3.15.3B) how the relationships between internal c-hulchul and external c-hulchul of a body are different when the body is an inner or outer body of m. We will call the two ranges of bodies as follows: an elementary body through the middle body as the **first half of the lineal range of bodies,** and the middle body through the U body as the **second half of the lineal range of bodies**.

Questions related to a middle body:

What is the absolute speed of a middle body?
Is it one half of the speed of light?
Is it one third of the speed of light?
Is I < C, I > C or I = C? (I = E for a middle body.)
Are there any examples of real physical bodies?
Could it be certain atoms?

3.15.4 Other Special Situations of C-Hulchuls

- **CH-Equivalent Bodies**: Two bodies will be called CH-equivalent if corresponding c-hulchuls are equal for each type of c-hulchul.

- **ICH-Equivalent Bodies**: Two bodies will be called ICH-equivalent if their internal c-hulchul is the same but the two bodies are not CH-equivalent.

In the case of ICH-equivalent bodies, composite c-hulchul decreases outer-body-ward.

Proof:

- Suppose b1 ⊂ b and ch(b1)= ch(b).
- Since ch{b1} = ch{b}, therefore,
  ch[[b1]] ∪ ch(b1)= ch[[b]] ∪ ch(b).
- Therefore, ch[[b1]] = ch[[b]], since ch(b1) = ch(b).
- Since b1 ⊂ b, therefore, ch[b1] ⊇ ch[b].
- Therefore, ch[[b1]] ∪ ch[(b1)] ⊇ ch[[b]] ∪ ch[(b)].
- Therefore, ch[(b1)] ⊇ ch[(b)] since ch[[b1]]= ch[[b]].
- **Therefore, composite c-hulchul decreases outer-body-ward in case of two ICH-equivalent bodies**.

- **CCH-Equivalent Bodies**: Two bodies will be called CCH-equivalent if their composite c-hulchul is same but the two bodies are not CH-equivalent.

- **ECH-Equivalent Bodies**: Two bodies will be called ECH-equivalent if their external c-hulchul is the same but the two bodies are not CH-equivalent.

Example: Consider a stationary car on the surface of the Earth. Since the car is stationary, therefore, the car and the Earth have the same external c-hulchul but their internal c-hulchul cannot be the same since the Earth is so much bigger than the car. See Table 3.15.7B for more examples. Moreover, if ch[[Car]] = ch[[Earth]], too, then we can prove the two bodies are CH-equivalent.

Suppose bodies b1 and b are ECH-equivalent bodies, where b1 ⊂ b. Then

$$ch[b1] = ch[b], \text{ that is, Ec is the same} \qquad (3.15.4A)$$

Property #1 of ECH-equivalent Bodies: (This is independent of Proposition #1.)

$$\mathbf{ch[(b1)] \subseteq ch[(b)]} \qquad \mathbf{(3.15.4B)}$$

(In general, composite c-ticks do not follow any inheritance property.)

Proof:

Let x ∈ ch[(b1)].

Therefore, x ∈ ch(b1), and x ∈ ch[b1] by the definition of ch[(b1)].

Therefore, x ∈ ch(b) because of the inheritance of internal c-ticks.

And x ∈ ch[b], in view of (3.15.4A).

Therefore, x ∈ ch[(b)].

This proves (3.15.4B).

Note: (3.15.4B) is true for the two ECH-equivalent bodies b1 and b whether Proposition #1 is true or not.

Property #2 of ECH-equivalent Bodies: (Assuming Proposition #1 is true.)

$$ch((b1)) = ch((b)), \text{ that is, I is the same} \qquad (3.15.4C)$$

Proof:

Since Proposition #1 is true, therefore

$$ch\{b1\} = ch\{b\} \qquad\qquad (3.15.4D)$$

(3.14.4D) implies

$$ch[b1] \cup ch((b1)) = ch[b] \cup ch((b))$$

This proves (3.15.4C), since external c-hulchul and pure internal c-hulchul of a body are disjoint, and in view of (3.15.4A).

Note: There is no additional property in the case of two ECH-equivalent bodies other than the inheritance properties, and (3.15.4B) and conditional (3.15.4C). Still, we have ch[[b1]] ⊇ ch[[b]] and not ch[[b1]] = ch[[b]]. **This means: Two ECH-equivalent lineal bodies b1 and b may have different time outage.**

Why do we have ch[[b1]] ⊇ ch[[b]] and not ch[[b1]] = ch[[b]] even if ch[b1] = ch[b]? (Though, in general, ch[[b1]] ⊇ ch[[b].) Consider the following diagram:

** Figure 3.15.4

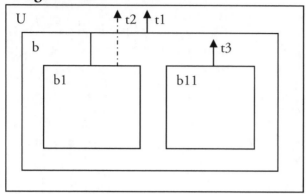

We note from the diagram:
b1 ⊂ b
b11 ⊂ b
t1 = t[b];
t3 = t[b11, b]
t1 = t2 = [b1], t2 is inherited from t1;
t[b1, b] is empty, and therefore, t[b1] = t[b] since t[b1] = t[b1, b] ∪ t[b].

- Now suppose t3 is concurrent with t1. As a result, t1 is not a pure external tick of b, t1 is a composite tick of b.
- But t1 is a pure external tick of b1, since t[b1, b] is empty and b1 may not have an internal tick concurrent with t1.
- We will have similar corresponding results in terms of c-ticks.
- This means b1 has an extra pure external c-tick compared to what b has.

In general, ch[[b]] increases inner-body-ward. This does not change even in case of two lineal ECH-equivalent bodies. However, we have ch[(b1)] ⊆ ch[(b)] in case of two ECH-equivalent bodies b1 and b; though, in general, composite c-hulchul does not follow any inheritance rule. Since

$$ch[b1] = ch[[b1]] ∪ ch[(b1)]$$
$$ch[b] = ch[[b]] ∪ ch[(b)]$$
$$ch[[b1]] ⊇ ch[[b]]$$
$$ch[(b1)] ⊆ ch[(b)],$$

it is possible that in some case corresponding values of pure external c-hulchul for b1 and b may be the same. However, in general, ch[[b1]] ⊇ ch[[b]] due to inheritance property of pure external c-hulchul.

3.15.6 The Second Equation in I, E and C is Untenable

(The first equation is the one from Proposition #1.)

This is because the presence of the second equation implies that if we know the value of one of I, E or C, then we can determine the value of the other two uniquely. This is contradicted by the example of the stationary car and the Earth having the same value of Ec but different values of Ic. See Table 3.15.7B for more examples.

3.15.7 External Motion in Terms of External Ticks

What represents external motion—external c-ticks, maximal external ticks or external xp-ticks? Since distance traveled is relevant to motion, we may also ask: What represents distance traveled—external c-ticks, maximal external ticks, or external xp-ticks? There is one-one correspondence between external c-ticks and external maximal ticks. A c-tick and a maximal external tick may involve more than one concurrent xp-tick. Below, we are presenting Table 3.15.7A showing external xp-ticks and external c-ticks (same as external maximal ticks) for three bodies b1, b and B where b1 ⊂ b ⊂ B, with all possibilities c-tick by c-tick.

Notes:

- Since b1 ⊂ b ⊂ B, therefore, b1 inherits external xp-ticks of b and B; b inherits external xp-ticks of B.

- Earlier, we assumed that external motion of a body B is the number of the maximal external ticks of B, which is the same as the number of external c-ticks. It is possible that the number of independent external xp-ticks forming a maximal external tick or c-tick have their own physical significance.

- Directions of ticks are immaterial.

- **Table 3.15.7A**: Column C2 shows all possible combinations of xp-ticks of b1, b and B. Column C3 shows corresponding counts of xp-ticks. Column C4 shows corresponding counts of c-ticks, which are the same as counts of external maximal ticks. In the table, c-ticks occur in the order of increasing number of independent xp-ticks because we wanted to make visually sure that all combinations of xp-ticks of the three bodies are there in the column C2; however, this need not be so. All ticks in the table are external ticks. **Numbers in the columns C3 and C4 are <u>not</u> cumulative vertically.**

- Classical motion is determined in terms of distance traveled in a certain direction. Here, we determine it as a number of c-ticks.

** Table 3.15.7A: (External motion versus external ticks)

C1	C2 p-ticks Y = tick N = no tick			C3 Number of Independent xp-ticks			C4 Number of c-ticks or maximal ticks		
c-tick #	b1	b	B	b1	b	B	b1	b	B
1	N	N	N	0	0	0	0	0	0
2	Y	N	N	1	0	0	1	0	0
3	N	Y	N	1	1	0	1	1	0
4	N	N	Y	1	1	1	1	1	1
5	Y	Y	N	2	1	0	1	1	0
6	Y	N	Y	2	1	1	1	1	1
7	N	Y	Y	2	2	1	1	1	1
8	Y	Y	Y	3	2	1	1	1	1

Example: How to compute five c-hulchul values Ic, I, C, E, Ec for a body b from the corresponding values for an inner body of b.

Here we will build a table of the five c-hulchul values that abide by the following conditions.

1. Bodies are shown in the first column in lineal order.
2. The values of I and Ic increase or remain the same outer-body-ward.
3. The values of E and Ec decrease or remain the same outer-body-ward.
4. The value of C may increase or decrease—true to its nature as it serves as a buffer between internal and external c-hulchuls.
5. I+C+E = 100 % of universal c-ticks.

** **Table 3.15.7B: Example with all c-hulchul values**

Body	Ic I+C	I	C	E	Ec E+C	ICE I+C+E
e	0	0	0	100	100	100
b1	10	3	7	90	97	100
b2	10	4	6	90	96	100
b3	22	12	10	78	88	100
b4	30	22	8	70	**78**	100
b5	32	22	10	68	**78**	100
b6	**55**	25	30	45	75	100
b7	**55**	27	28	45	73	100
b8	60	40	20	40	60	100
b9	80	50	30	20	50	100
U	100	100	0	0	0	100

Table 3.15.7B shows consistent theoretical values of c-hulchul for different lineal bodies to demonstrate the concept; this is not empirical data.

We will discuss this more in Chapter 4.

Notes

- **Bodies b4 and b5 are ECH-equivalent**: Bodies b4 and b5 have same value of their external c-hulchuls but different values of their internal c-hulchuls. The corresponding values are highlighted in the table. Note that the value of C increases and the value of E decreases from body b4 to body b5.

- **Bodies b6 and b7 are ICH-equivalent**: Bodies b6 and b7 have the same value of their internal c-hulchuls but different values of their external c-hulchuls. The corresponding values are highlighted in the table. Note that value of C decreases from body b6 to body b7.

CHAPTER 4

Estimation of Composite Concurrency Hulchul

4.1 Estimation of Composite Concurrency Hulchul

As already discussed in Chapter 3, the composite concurrency hulchul (c-hulchul) of a body b is the set of composite concurrency ticks (c-ticks) of b. A composite c-tick of b is both an internal and external c-tick of b. (Same p-tick cannot be both an internal and external p-tick of the same body b. Same c-tick can be both an internal and external c-tick of the same body b because if an internal p-tick and an external p-tick of b are concurrent, then the two p-ticks belong to the same c-tick.)

$$ch[(b)] = CINT(ch(b), ch[b]) = CINT(CNCY(xph(b)), CINT(xph[b]).$$

4.1.1 Essence of Composite Ticks/C-Ticks

A body b will have a composite c-tick if and only if b has a composite tick; a composite tick of b, in turn, arises because both b and one of its inner or outer bodies have external prime ticks concurrently and independently.

4.1.2 Review of Previous Chapters

In this chapter, we will use some of the concepts and their properties established in the previous chapters; the specific concepts and their properties, we will use, are:

Extended Prime Hulchuls (xp-hulchuls)
- xph(b) – Internal xp-hulchul of the body b
- xph[b] – External xp-hulchul of the body b
- xph[b, B] – Relative xp-hulchul of b in B where b ⊂ B
- xph((b)) – Pure internal xp-hulchul of the body b
- xph[[b]] – Pure external xp-hulchul of the body b
- xph[I(b)] – Composite internal xp-hulchul of the body b
- xph[E(b)] – Composite external xp-hulchul of the body b
- xph{b} – Total xp-hulchul of the body b

Concurrency Hulchuls (C-Hulchuls)

- ch(b) = CNCY(xph(b)) – Internal c-hulchul of the body b
- ch[b] = CNCY(xph[b]) – External c-hulchul of the body b
- ch[b, B] = CNCY(xph[b, B]) – Relative c-hulchul of b in B where b ⊂ B
- ch((b)) = CNCY(xph((b))) – Pure internal c-hulchul of the body b
- ch[[b]] = CNCY(xph[[b]]) – Pure external c-hulchul of the body b
- ch[(b)] = CNCY(xph[I(b)]) = CNCY(xph[E(b)] – Composite c-hulchul of the body b
- ch{b} = CNCY(xph{b}) – Total c-hulchul of the body b
- ch(U) = Universal c-hulchuls; its members are called universal c-ticks and are used as instances of time mathematically

Counts of C-Ticks in C-Hulchuls

Count of members in a c-hulchul is denoted by replacing "ch" by "CH" in the name of the c-hulchul. For example, count of ch(b) is denoted by CH(b).

A shorter notation for counts of some of the above sets (with assumed underlying body b) is:
- Ic = CH(b)
- Ec = CH[b]
- I = CH((b))
- E = CH[[b]]
- C = CH[(b)]
- ICE = CH{b}

- ICE2 = CH(b) + CH[b]
- Ic = I + C = CH((b)) + CH[(b)] = CH(b)
- Ec = E + C = CH[[b]] + CH[(b)] = CH[b]
- ICE = I + C + E = CH((b)) + CH[(b)] + CH[[b]]
  = Ic + E = CH(b) + CH[[b]]
  = Ec + I = CH[b] + CH((b))
  = CH(U)
- ICE2 = Ic + Ec = I + 2C + E = CH(b) + CH[b]

Some of the Properties of C-hulchuls

- ch(b) = ch((b)) ∪ ch[(b)]
- ch[b] = ch[[b]] ∪ ch[(b)]
- ch[b] = ch[b, B] ∪ ch[B] where b ⊂ B
- ch{b} = ch((b)) ∪ ch[(b)] ∪ ch[[b]]
  = ch[b] ∪ ch((b))
  = ch(b) ∪ ch[[b]]

- **Proposition #1: Total c-hulchul is constant**: Total c-hulchul of all bodies is the same for the same h-domain; that is, I+C+E is the same for all bodies for the same h-domain.
- **Inheritance of internal c-hulchul**: Internal c-hulchul values ch(b), ch((b)), CH(b) and CH((b)) remain the same or increase outer-body-ward.
- **Inheritance of external c-hulchul**: External c-hulchul values ch[b], ch[[b]], CH[b] and CH[[b]] remain the same or increase inner-body-ward.
- **No rule for inheritance of composite c-hulchul**: There is no inheritance rule in case of the composite c-hulchul ch[(b)] or its count CH[(b)]. This makes it more difficult to estimate the composite c-hulchul.
- **Uniform c-hulchuls in large h-domains**: Distribution of each of the three types of c-ticks—pure internal c-ticks, pure external c-ticks and composite c-ticks is statistically uniform in all large h-domains for a body.
- **Time Outage:** Time outage is the reason for time dilation. Time outage occurs during pure external c-hulchul, ch[[b]], intermittently for a common body b.

Our **goal** in this chapter is to analyze c-hulchul to be able to answer the following questions and justify the answers:

1. Does each common body have some composite ticks/c-ticks? (**Logically, we are not yet certain.**)

2. Can there be a body having only composite c-ticks (that is, there are no pure c-ticks)? (**Statistically, no.**)

3. Can we come with a general rule, in terms of an equation or inequality, to estimate the number of composite c-ticks correctly? (**Inequality? Yes.**)

4. Can we at least determine the minimum and maximum limits on the number of composite c-ticks of a body? (**Maximum limit? Yes.**)

4.2 Second Equation of Hulchul in I, E and C is Untenable

After determining the first equation of hulchul in I, C and E, associated with Proposition #1on c-hulchul, that I+C+E is 100% of CH(U) for all bodies for the same h-domain; initially, we had thought we could find another independent equation of c-hulchul in I, C and E so the two equations could lead to unique values of I, C and E, if we knew the value of one of I, C or E. Then we found an example that contradicts this assumption.

Consequence of having two independent equations in I, C and E and C: Suppose we have two independent equations in the three variables I, C and E. If values of one of the three variables, say I, are the same for two different bodies, then the values of E must also be the same, and the values of C must also be the same for the two bodies.

We know now show that, in fact, there exists an example of two bodies having the same values of I but different corresponding values of E and different corresponding values of C. This makes, in general, a second equation in I, E and C untenable.

Example: Two bodies having the same values of I but different corresponding values of E and C:

Consider a stationary car on the surface of the Earth. Since the car is at rest relative to the Earth, the car and the Earth have the same external c-hulchul. On the other hand, they have different internal c-hulchuls since the Earth is so much bigger than a car; if there is any doubt about the Earth having more internal c-hulchul than the car has, then we can choose a much smaller body than a car, say a stationary dust particle, to make the difference between the internal c-hulchuls of the two bodies more pronounced. This means the stationary car and the Earth have the same values of I but different values of E. (Also see section 3.15.4.)

Example: Two bodies having the same internal c-hulchul but different external c-Hulchul

It should be possible though we could not yet identify an example.

4.3 Guiding Principles on Composite C-hulchul

4.3.1 Based on Properties Composite Ticks/C-ticks

In Chapter 3, we observed that I and Ic increase outer-body-ward, E and Ec decrease outer-body-ward **but C does not exhibit this behavior**. Still, some other properties of c-hulchul do have a significant bearing on composite ticks and c-ticks. With their help we will attempt to estimate the value of C. We will list all such important properties here. Many of these properties apply to both composite ticks and composite c-ticks.

Note: In some of the examples below, we will use sequences of X's and Y's to represent a mix of c-ticks where an X represents an internal c-tick and a Y represents an external c-tick.

1. **Why composite ticks/c-ticks are necessary:** Without composite ticks, dynamics of bodies will become slow, even very slow; as in that case, before a common body can tick, it must ensure that it does not tick concurrently and independently with any of its inner or outer body.

2. **Axiom of concurrent and independent external ticks implies existence of composite ticks**: This means without composite ticks, a body cannot have two external ticks concurrently and independently. This is too severe a restriction. Example: If a person moves in a train and the train also moves and the two motions are concurrent and independent then this implies a composite motion (and composite ticks).

   Maximal external ticks and chained ticks: All outer bodies of b in a maximal external tick of b are composite tick in case length of the maximal external tick is greater than 1. **Without composite ticks, we will not have maximal external ticks or even chained ticks.**

3. **Only a common body can have composite ticks/c-ticks:** The U body and elementary bodies cannot have composite ticks/c-ticks.

4. **Maximum value of C ≤ minimum(Ic, Ec).** This is because C cannot exceed either of Ic and Ec.

5. **Pure C-ticks and Composite C-ticks are mutually exclusive:** It takes two consecutive, concurrent c-ticks of different types to form a composite c-tick. If a c-tick is not a composite c-tick then it is a pure c-tick. If a c-tick is not a pure c-tick then it must be one the two consecutive c-ticks of different types that form a composite c-tick.

6. **Consecutive c-ticks of the same type: Two consecutive c-ticks of the same type cannot be concurrent.** If this were so then two c-ticks would have been combined into one c-tick initially. In the sequence of c-ticks $Y_1X_2X_3Y_4$ either each of Y_1X_2 and X_3Y_4 forms a composite c-ticks or at least one of X_2 and X_3 is a pure c-tick. **Three consecutive c-ticks of the same type imply that the middle c-tick must be a pure c-tick.**

7. **Values of I and E are greater than zero% and the value of C cannot be 100% of the total c-hulchul in sufficiently large h-domains for common bodies:** This is based on the observation that the middle X in XXX must be a pure c-tick. Similarly, the middle Y in YYY must be a pure external tick. Probabilistically, the number of triplets of X's and Y's is greater than zero in sufficiently large h-domains. This implies that a common body in a sufficiently large h-domain has some pure internal c-ticks and some pure external c-ticks and therefore all c-ticks cannot be composite c-ticks. (At times, even two consecutive c-ticks of the same type may result in a pure c-tick. Examples: in XXX the middle X is certainly a pure internal c-tick. In $X_1Y_2X_3X_4Y_5$, if X_1Y_2 is a composite c-tick then X_3 must be a pure c-tick.

8. **Uniform distribution of composite c-ticks**: The pattern of distribution of c-ticks of different types should be uniform h-domain-wise in case of uniform c-hulchul for a body. This means: if S is a sufficiently large sequence of c-ticks for a common body b, then S has a definite pattern of distribution of c-ticks of the three types. In that case, any super-sequence of S or any large subsequence of S should have the same pattern of distribution of c-ticks of the three types as S has. Therefore, distribution of composite c-ticks, in particular, should also be uniform h-domain-wise. For example: suppose a body b has 30% internal c-ticks and 70% external c-ticks. Consider the sequence S1 of c-ticks in an h-domain h1: XXXYYYYYY and suppose each X represents 100 consecutive internal c-ticks and each Y represents 1 million external c-ticks. (The sequence S1 can yield at most one composite c-tick.) The sequence S1 is highly unlikely to occur because S1 has a separate high concentration of X's and a separate high concentration of Y's for 30% internal c-ticks and 70% external c-ticks. Moreover, an h-domain larger than h1 of the body b will retain the sequence S1 as a subsequence violating the principle of uniformity of c-hulchul h-domain-wise as discussed in the previous chapter.

9. **The second equation of hulchul in I, E and C is untenable:** This is as discussed earlier in the chapter.

4.3.2 Other Properties of Composite Ticks/C-Ticks

10. **A pure c-tick is inherited as a pure c-tick only:** A pure c-tick, of either type, is inherited as a pure c-tick of the same type.

11. **Limited inheritance of composite c-ticks**: Composite c-ticks in general are not inherited. However, they may be inherited in special cases. For example: Suppose b ⊂ B, b has an internal tick t1 and B has an external tick t2 concurrently. Then each body b0, where b ⊂ b0 ⊂ B, has a composite c-tick, since b0 inherits t1 as an internal tick and t2 as an external tick.

12. **From one body to another body, types of two ticks may change but their concurrency or order does not change:** A tick of one type of a body may be of another type for another body. Two internal ticks may form a composite tick when inherited by another body and vice versa. Similarly, two external ticks may form a composite tick when inherited by another body and vice versa. Example:

    Suppose $b1 \subset b \subset B$, t1 = t[b1, b], t2 = t[b, B].
    t3 = t1 + t2 = t[b1, B].
    t3 is inherited partly from t1 as an internal tick of B.
    t3 is also inherited partly from t2 as an external tick of b1.
    So t1, t2 and t3 are external ticks of b1.
    Also, t1, t2 and t3 are internal ticks of B.

t1 and t2 form a composite tick of b; both t1 and t2 are concurrent external ticks of b1 and any inner bodies of b1; both t1 and t2 are concurrent internal ticks of B and any outer bodies of B.

If t1 and t2 are not concurrent and t1 occurs before t2 then it is true with respect to any body having t1 and t2 as ticks. In particular, two consecutive ticks remain consecutive in the same order.

** **Figure 4.3.2**

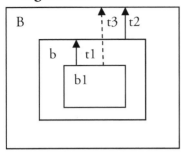

13. **Geometric representation of c-hulchuls:** Ic, Ec and ICE for a body can be represented by the three sides of an obtuse angled triangle. In this form, they still abide by Proposition #1and inheritance properties of c-hulchuls: Ic, Ec, I and E. ICE corresponds to the largest side of the triangle; Ic and Ec correspond to the other two sides. I and E are determined by subtracting Ic and Ec respectively from ICE. So the side corresponding to Ic can be divided into I and C; the side corresponding to Ec can be divided into E and C; the side corresponding to ICE can be divided into I, C and E. Also, in that case, algebraically, the relation between I, E and C can be represented by:

$$(I+C)^2 + (E+C)^2 - 2(I+C)(E+C)COS(A) = (I+C+E)^2$$

Where A is the angle opposite to the side represented by ICE.

Note: We do not yet know if there is any significance of this geometric representation; it is simply casual as of now.

14. **The two halves of a composite tick**: A composite c-tick needs two distinct c-ticks, one internal c-tick and one external c-tick, with the two c-ticks being concurrent. An internal c-tick and an external c-tick forming a composite c-tick are like two halves of a composite c-tick.

15. **Question**: Suppose a body has 50% internal c-ticks and 50% external c-tick in an h-domain and the corresponding percentages

for another body are 30 and 70. Which body is likely to have more composite c-ticks?

To see the upper limits on C in the two cases, see Table 4.6.1B and Table 4.6.2B. Probabilistically, the upper limit in the 50-50 case is higher than in the 30-70 case.

4.3.3 Relative Change in Different C-hulchul Values

We will briefly discuss here how change in values of some of the c-hulchul variables, I, C, E, Ic and Ec, may affect the change in other variables.

1. **If Ic does not change then E does not change and vice versa**
    since $I+C+E = Ic + E = Ec + I$ and the value of $I + C + E$ is constant. If either Ic or E increases then the other decreases.

    Similarly, if Ec does not change then I does not change and vice versa. If either Ec or I increases then the other decreases.

2. **Change in Ic $\geq$ Change in C outer-body-ward.**
    Proof: We know $\Delta I \geq 0$ outer-body-ward.
    Therefore, $\Delta I + \Delta C \geq \Delta C$ outer-body-ward.
    That is, $\Delta Ic \geq \Delta C$ outer-body-ward

3. **Change in Ec $\geq$ Change in C inner-body-ward.**
    Proof: We know $\Delta E \geq 0$ inner-body-ward.
    Therefore $\Delta E + \Delta C \geq \Delta C$ inner-body-ward.
    That is, $\Delta Ic \geq \Delta C$ inner-body-ward.

4. **Identity: IcEc $-$ IE = C(I+C+E).** This can be readily verified. It should be noted here how C is directly related to the difference IcEc $-$ IE, remembering that $I+C+E$ is constant.

4.4 Initial Thoughts on Deriving Time Dilation from C-Hulchul

In the middle stage of this research, after we established Proposition #1, we intuitively thought of the following as the other equation that could help us determine the time dilation in terms of c-hulchul:

$$CH^2(b) + CH^2[b] = CH^2(U) \tag{4.4A}$$
$$\text{Or} \quad (I+C)^2 + (E+C)^2 = (I+C+E)^2 \tag{4.4B}$$
$$\text{Or} \quad C^2 = 2IE \tag{4.4C}$$

Now, for a given h-domain and body b, let:

u = Number of universal c-ticks per unit of time
Tu = Number of units of time from the universal c-hulchul
Tu is the universal time
v = Number of external c-ticks of b per unit of time
Tb = Number of units of time from the internal c-hulchul of b
Tb is the time for the body

Therefore,

$$CH(U) = u^*Tu$$
$$CH(b) = u^*Tb$$
$$CH[b] = v^*Tu$$

Therefore (4.4A) can be written as
$$(u^*Tb)^2 + (v^*Tu)^2 = (u^*Tu)^2$$
$$\text{Or} \quad (u^*Tb)^2 = (Tu)^2(u^2 - v^2)$$
$$\text{Or } (\quad Tb)^2u^2 = (Tu)^2(u^2 - v^2)$$

$$\text{Or} \quad \frac{Tu}{Tb} = \frac{1}{\sqrt{1 - \dfrac{v^2}{u^2}}} \tag{4.4D}$$

The R.H.S of the equation (4.4D) is very similar to the time dilation factor in the Theory of Relativity. The above conclusion is based on the equation (4.4A) but the very presence of another equation in I, C and E, aside from the first equation of hulchul, violates the guiding principle #8, though the procedure is correct. Still, the equations (4.4A) and (4.4D) may be true for a specific body or for a class of specific bodies, but they are not true in general.

Note: Why is ch[b] = v*Tu and not ch[b] = v*Tb? This is because we are using ch[b], the absolute external c-hulchul of the body b. Therefore, we should use absolute time that is Tu and not Tb. In case, we use a relative external c-hulchul, then Tb may be considered. However, we are not sure which time we should use since there can be more than one relative external c-hulchuls. Both Tb and Tu are unique—associated with the body b and the body U respectively.

In any case, in general, we cannot have equations (4.4A) or (4.4B) as we have already demonstrated that a **second hulchul equation in I, C and E is untenable**.

There is another difficulty with (4.4D); Tu / Tb, depends on internal c-hul of a body which is not the same two ECH-equivalent bodies. Time dilation must be the same for two ECH-equivalent bodies since they have the same external c-hulchul.

4.5 Difficulty with Finding the Number of Composite C-Ticks

At the current stage of our research, we are unable to determine the value of C precisely, that is, the number of composite c-ticks of a body b, in a given h-domain; however, we can determine an upper limit on the value of C as we will show in this section. Consequently, we will also be able to determine lower limits on the values of I and E, the numbers of pure internal c-ticks and pure external c-ticks respectively, since I+C+E has the same value for all bodies in a given h-domain. For this purpose, we need to introduce the following concept that is less restricted than a composite c-tick is.

4.5.1 Composite Concurrency Pairs (Composite C-pairs) as Potential Composite C-ticks

Definition: An internal c-tick X and an external c-tick Y of a body b make a composite c-pair of b if and only if they meet each of the following two conditions:

1. X and Y are consecutive c-ticks of b.

2. Any two composite c-pairs of b are always disjoint; that is, two composite c-pairs cannot share a c-tick between them.

For example: Suppose $X_1Y_2X_3Y_4$ is a sequence of consecutive c-ticks. Both X_1Y_2 and Y_2X_3 cannot be composite c-pairs as X_1Y_2 and Y_2X_3 are not disjoint. But both X_1Y_2 and X_3Y_4 can be composite c-pairs.

The difference between a composite c-tick and a composite c-pair: A composite c-tick is always a composite c-pair but a composite c-pair may or may not be a composite c-tick because the two c-ticks in a composite c-tick are always concurrent but the two c-ticks in a composite c-pair may or may not be concurrent; this is precisely the difference between a composite c-tick and a composite c-pair.

Thus, the conditions that make a composite c-pair are necessary to make a composite c-tick but they are not sufficient.

Uncertainty on concurrency of two c-ticks of different types: At the current stage of this research, we have no direct handle on figuring out logically or theoretically whether two consecutive c-ticks of different types are concurrent or not. Therefore, we cannot precisely determine the number of composite c-ticks of a body in a given h-domain. However, we can determine the number of composite c-pairs of a body correctly as we will see in the following paragraphs. The number of composite c-pairs can serve as the upper limit on the number of composite c-ticks.

Married couples—an example similar to composite c-ticks: Suppose 100 adult men and women, in the same age group, are standing in a line. What is the probable number of married couples in the line? It is reasonable to assume that a man and woman who are married are likely to be together in the line. However, a man and a woman together in the line may not necessarily be a married couple; this is very similar to the situation in the case of composite c-pairs. However, the similarity ends here. Whether a man and a woman standing together in the line are a married couple or not may be determined by other well-known means—such as asking them directly. However, we have no such means at our disposal, at this stage, to determine whether two consecutive c-ticks of different types are concurrent or not. (We cannot put this question to c-ticks!) However, it is possible that some means do exist in nature to figure out whether two consecutive c-ticks of different type are concurrent or not.

4.6 Number of Composite C-Pairs as Limits on Composite C-Hulchul Values

Now we will develop two different methods to estimate the number of c-pairs. The two methods are: the **Probabilistic Method** and the **Simulation Method**. Whereas the simulation method is more assured, it does not lead to a mathematical expression (formula) but it does lead to actual numbers. The probabilistic method is slightly less assured but it does leads to a mathematical expression. We will discuss both methods here; surprisingly, the outputs of the two methods are very close to each other.

4.6.1 Probabilistic Method to Find Limits on C-Hulchul Values

Let us consider a sufficiently long sequence S of c-ticks of the body b: We will assume S is a random mix of internal and external c-ticks. Suppose the number of internal c-ticks and external c-ticks in S are Ic and Ec respectively. We will now estimate the probable number of composite c-pairs in S. We will denote each internal c-tick by an X and each external c-tick by a Y in S. Here, we will assume an infinite extension of the sequence S exists though we will use only a finite part of it.

The number of all possible pairs of the type XX is: $(\mathbf{Ic})^2$

The number of all possible pairs of the type YY is: $(Ec)^2$
The number of all possible pairs of the type XY or YX is: **2IcEc**
The number of all possible composite c-pairs is: **IcEc**

(Not 2IcEc because composite c-pairs must be disjoint and as a result both X_1Y_2 and Y_2X_1 cannot be composite c-pairs.)

Therefore, for Ic+ Ec c-ticks in S:

The total number of c-ticks is: **Ic + Ec**
The total number of possible pairs of c-ticks is: $(Ic)^2 + (Ec)^2 + IcEc$
The probability of a composite c-pair is:

$$\frac{IcEc}{(Ic)^2 + (Ec)^2 + IcEc}$$

The probable number of composite c-pairs for a total of Ic + Ec c – ticks is:

$$\frac{IcEc(Ic + Ec)}{(Ic)^2 + (Ec)^2 + IcEc}$$

Since a composite c-pair may or may not be a composite c-tick, therefore, the number of composite c-pairs is the upper limit on the number of composite c-ticks. So:

$$\textbf{\textit{Maximum value of C }} \leq \frac{IcEc(Ic + Ec)}{(Ic)^2 + (Ec)^2 + IcEc} \qquad \textbf{(4.6A)}$$

Since I = Ic – C and E = Ec – C, therefore:

$$\textbf{\textit{Minimum value of I }} \geq \frac{(Ic)^3}{(Ic)^2 + (Ec)^2 + IcEc} \qquad \textbf{(4.6B)}$$

$$\textbf{\textit{Minimum value of E}} \geq \frac{(Ec)^3}{(Ic)^2 + (Ec)^2 + IcEc} \qquad (4.6C)$$

(4.6B) and (4.6C) can be readily verified using the value of C as given by (4.6A).

Note: The three above inequalities (4.6A), (4.6.B) and (4.6C) do not provide exact values of I, C and E; they just set upper limit on C and lower limits on I and E. This is because, the three inequalities were arrived at using **c-pairs** (consecutive pairs of an internal tick and an external c-tick) and not **composite c-ticks**. A c-pair may or may not be a composite c-tick. If a c-pair is not a composite c-tick, then the actual value of C is less by one, and the actual values of I and E each is more by one. In some situations, this logic may be more complex as it may affect the neighboring c-ticks; for example: $X_1Y_2X_3X_4$. In this case, X_1Y_2 is a c-pair and X_3 is pure internal c-tick and therefore, C = 1, I = 1 and E = 0. However, if X_1Y_2 is not a c-pair then Y_2X_3 is c-pair. Therefore, again, C = 1, I = 1 and E = 0, since, now, Y_2X_3 is a c-pair. But in the case of $Y_2X_1X_3X_4$, if Y_2X_1 is not a c-pair, then the values of I, C and E are affected.

A casual geometric interpretation of the expression for the denominator $(Ic)^2 + (Ec)^2 + IcEc$: If we represent the two sides of a triangle by Ic and Ec and the angle between them is 120° then the third side is precisely the value of the denominator.

A simplified view of the limitations on I, E and C

1. **C ≤ 2Ic/3**
   This follows from (4.6A) and assuming C has maximum value when Ic = Ec. Replace Ec by Ic in (4.6A).
2. **C ≤ 2Ec/3**
   As above.
3. **C ≤ ICE/2**
   Add 1) and 2).
   6C ≤ 2(I+C) + 2(E+C) = 2(I+C+E) + 2C
   Or 2C ≤ ICE or C ≤ ICE/2.

4. **C ≤ I+E**

   3) implies: $2C \leq I+C+E$ or $C \leq I+E$.

5. **C ≤ 2I**

   1) implies: $3C \leq 2(I+C)$ or $C \leq 2I$.

6. **C ≤ 2E**

   This follows from 2) as above.

7. **Ic/3 ≤ I ≤ Ic**

   Using 5): $C \leq 2I$ or $Ic \leq 3I$ or $Ic/3 \leq I$. We already know $I \leq Ic$.

8. **Ec/3 ≤ E ≤ Ec**

   Using 6) and as above.

Graphic View of the Relative C-hulchul Values

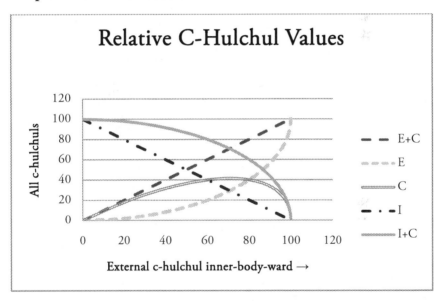

Notes:

1. The purpose of the above graphic representation is just to demonstrate a pattern of the relative values of different c-hulchuls.
2. The values are shown as percentages of total c-hulchul ICE.
3. Values of Ec have a straight line curve because they are also used as x-axis (reference values).
4. Values of I have a straight line curve because Ec + I is constant.

5. Intersections of Ic and Ec, intersection of E and I, and point of maximum value on C have the same values; they represent the middle body.

Computation of Probabilistic Values of C-hulchuls

In the Table 4.6.1A below, we have computed the limits on the values of I, E and C, as implied by (4.6A), (4.6B) and (4.6C), for different combinations of the values of Ic and Ec as shown in the columns C3, C4 and C5 respectively:

** **Table 4.6.1A: Probabilistic Hulchul Values with ICE2 = 100**

C1	C2	C3	C4	C5	C6	C7
Ic%	Ec%	Minimum I	Minimum E	Maximum C	ICE I+C+E	ICE2 Ic + Ec
0	100	0	100	0	100	100
2	98	0.0008	96.0008	1.9992	98.0008	100
10	90	0.1099	80.1099	9.8901	90.1099	100
20	80	0.9524	60.9524	19.0476	80.9524	100
30	70	3.4177	43.4177	26.5823	73.4177	100
40	60	8.4211	28.4211	31.5789	68.4211	100
50	50	16.6667	16.6667	33.3333	66.6667	100
60	40	28.4211	8.4211	31.5789	68.4211	100
70	30	43.4177	3.4177	26.5823	73.4177	100
80	20	60.9524	0.9524	19.0476	80.9524	100
90	10	80.1099	0.1099	9.8901	90.1099	100
98	2	96.0008	0.0008	1.9992	98.0008	100
100	0	100	0	0	100	100

Notes on Table 4.6.1A

1. **Order of information**: The table shows the information for a number of bodies, one row per body, outer-body-ward.
2. **An h-domain is assumed**: The table shows the information in different rows for the same h-domain.

3. **How we chose the values for columns C1 and C2**: Each body has definite values of internal c-hulchul Ic and external c-hulchul Ec in the given h-domain; Ic and Ec are in a definite ratio. We chose the ratio so the sum of the two numbers involved in the ratio is 100. It is mentally convenient to choose the two numbers in the ratio so that their sum is 100 and the values in columns C1 and C2 are percentages of Ic and Ec to ICE2. Here, the ratio of Ic and Ec is important; the actual values of Ic and Ec, which may run into astronomical numbers, are immaterial as we assume values of c-hulchul are uniform h-domain-wise.

4. The values in columns C3, C4 and C5 are not actual values of I, E and C; they are limits on actual values.

5. Values in the row for the pair (0,100) correspond to an elementary body and the values in the row for the pair (100,0) correspond to the U body.

6. Note the symmetry between the values of certain rows. For examples: Symmetry between rows corresponding to the pairs (2,98) and (98,2), (10,90) and (90,10), and others.

7. Values in columns C1, C2, C3 and C4 follow the inheritance rule but the column C5, representing composite c-hulchul, does not follows that rule—this is characteristic of the composite c-hulchul.

8. **The value of I+C+E must remain the same** but it is not so in column C6; we will now correct it as explained below:

To make the values in different rows in the Table 4.6.1A comparable, we must change the value ICE to the same value and adjust the values of I, E and C accordingly; this is because ICE is the same for all bodies for the same h-domain. We chose ICE = 100 for the sake of mental convenience. Therefore, we will derive Table 4.6.1B from Table 4.1.6A as follows: In each row of Table 4.6.1A, we multiply I, E, C by 100 and divide by the corresponding value of ICE. This is permissible since the value of ICE is the same for all bodies for a given h-domain. C-hulchul values of different bodies in the Table 4.6.1B are now comparable. Note: since the values are symmetric for the pair (40,60) and (60,40), and others, we are not showing the rows for (60,40), (70,30), (80,20), (90,10), (98,2) and (100,0) in Table 4.6.1B.

** Table 4.6.1B: Probabilistic Hulchul Values with ICE = 100

C1	C2	C3	C4	C5	C6	C7
Ic	Ec	Min. I	Min. E	Max. C	ICE I+C+E	ICE2 Ic + Ec
0	100	0	100	0	100	100
2.0408 2%	99.9992 98%	0.0008	97.9592	2.0400	100	102.0400
11.0976 10%	99.8780 90%	0.1220	88.9024	10.9756	100	110.9756
24.7059 20%	92.8235 80%	1.1765	75.2941	23.5294	100	123.5924
40.8621 30%	95.3448 70%	4.6551	59.1379	36.2069	100	136.2069
58.4615 40%	87.6923 60%	12.3077	41.5385	46.1538	100	146.1538
75.0000 50%	75.0000 50%	25.0000	25.0000	50.0000	100	150.000

Notes on Table 4.6.1B:

1. In column C1, the percentages in the first pair are percentages to ICE2 and the percentages in the second pair are percentages to ICE.
2. Values in columns C1, C2, C3 and C4 follow the inheritance rule; as expected, values in C5 do not. Values in column C6 now follow the rule of Proposition #1 that the values of I+C+E remains the same for all bodies for the same h-domain.
3. The values of Ic and Ec percentages in columns C1 and C2 have percentages to both ICE and ICE2. The values in columns C3, C4 and C5 and ICE2 in Table 4.6.1B are different from the corresponding values in Table 4.6.1A but the values in columns C1 and C2 as percentages do not change in Table 4.6.1B as they are still the same percentages of the new values of ICE2. The values of ICE2 are shown in the column C7 in both tables.
4. Note that ICE2 = ICE + C in each row as it should be.

Significance of the Minimum Values of I (pure internal c-hulchul) and E (pure external c-hulchul)

In Table 4.6.1B, we show that for a common body and in sufficiently large h-domains:

- I > 0%
- E > 0%
- Ic < 100% since I + Ec = I+C+E = 100 and I > 0.
- Ec < 100% since E + Ec = I+C+E = 100 and E > 0.
- C < 100% since I+E > 0 and I+C+E = 100.

That is,

- 0% < I ≤ Ic < 100 %
- 0% < E ≤ Ec < 100 %

The first two bullets above are in conformity with guiding principle #7 on composite c-hulchul.

4.6.2 Simulation Method to Find Limits on C-hulchul Values

Here, we simulate occurrences of internal and external c-ticks of a body by generating letters X's and Y's repeatedly where an X represents an internal c-tick and a Y represents an external c-tick. The letters X's and Y's are generated randomly so the number of X's and number of Y's are generated in a certain ratio which is the same as the ratio Ic to Ec, which varies from body to body. The basic algorithm of **simulation method** is as follows:

Algorithm for Simulation Method

Repeat the procedure Simulate-Ratio-Ic-Ec as many-times as desired:

Simulate-ratio-Ic-Ec-Choice

- Choose two numbers Xc and Yc for the ratio Ic:Ec so Xc + Yc = 100.
- Generate the **first letter** randomly so that X's and Y's occur in the ratio Xc:Yc.
- Repeat Simulate-C-Ticks-Generation as many times as desired.

Simulate-C-Ticks-Generation

- Generate the **second letter** randomly so that X's and Y's occur in the ratio Xc:Yc.
- Compare the first and second letters.
- Both letters are X: Then the first letter represents a pure internal c-tick and the second letter is used again as the first letter in the next repetition of this process.
- Both letters are Y: Then the first letter represents a pure external c-tick and the second letter is used as the first letter again in the next repetition of this process.
- The two letters are different: Then the two letters together represent a composite c-pair. Neither of the two letters is used again in the process. Generate the first letter randomly so that X's and Y's occur in the ratio Xc:Yc. The letter generated here is used as the first letter again in the next repetition of this process.

End-of-Simulate-C-Ticks

End-of-Simulate-Ratio-Ic-Ec

For example, see the output of Simulate-C-Ticks-Generation below: The first line is a sequence as a mix of X's and Y's. The second line marks corresponding X's and Y's as pure internal c-ticks, pure external c-ticks and composite c-pairs by letters I, E and C respectively:

```
XXXXYYXYYYYYXYXYYYYYXXYXYYXXYXXYYYYX
IIIC C EEEEC C EEEEC C C C C IC EEC
```

ICE2 = Number of letters in the first line
ICE = Number of letters in the second line

Note: The use of I, E and C, above, should not be confused with the other use of I, E and C where they represent counts of ch((b)), ch[[b]], and ch[(b)] respectively.

We will now use the above algorithm to generate a sequence of internal c-ticks and external c-ticks in a given ratio Xc:Yc as a random mix of letters X's and Y's using random numbers until we have ICE = 2,000,000 and compute values of c-hulchuls as follows: (We assume here Xc+Yc = 100 and Ic:Ec = Xc:Yc.)

1. **Define the following variables and initialize them as shown**:
   - ICE = 2,000,000 (This is the size of the sequence of c-ticks we chose; any value, preferably a large value, can be chosen here.)
   - X-Count (to keep track of how many letters X's are generated) = 0
   - Y-Count (to keep track of how many letters Y's are generated) = 0
   - ICE2 = X-Count + Y-Count = 0
   - Minimum I = Pure-Internal-C-Ticks-Count = 0
   - Minimum E = Pure-External-C-Ticks-Count = 0
   - Maximum C = Composite-C-Pairs-Count = 0

2. **How to generate letter X or Y randomly**: Generate a random number and divide it by 100. If the remainder < Xc then we generate a letter X else we generate a letter Y and we **increment X-Count or Y-Count by 1** accordingly.
   Pairs (Xc, Yc) can have any values but it is easier mentally if their sum is 100. We choose the following pairs of values: (2,98), (10,90), (20,80), (30,70), (40,60), (50,50). Pairs (20,80) and (80,20), for example, will return symmetrical outputs.

   Use the following procedure:

3. Generate the very first letter of the sequence S as described in the step #2 above. This is the first letter of the two letters to be compared.

4. Do the following ICE (= 2,000,000) times:

- Generate the next letter of the sequence S as described in step 2 above. This is the second letter of the two letters to be compared.
- Compare the last two letters of the sequence generated and update the counts as follows:
- If each of the two letters is X then the first letter corresponds to a pure internal c-tick; accordingly, **increment Pure-Internal-C-Ticks-Count by 1.**
- If each of the two letters is Y then the first letter corresponds to a pure external c-tick; accordingly, **increment Pure-External-C-Ticks-Count by 1**.
- If the two letters are not equal then the two letters make a composite c-pair; accordingly, **increment Composite-C-Pairs-Count by 1, generate a new letter randomly,** as described in step #2 above, to be used as the first letter of the two letters to be compared.

5. Compute ICE2 = X-count+ Y-count.

6. Build the first Table 4.6.2A with rows corresponding to each pair of values of Xc and Yc having the following information:
Ic (= Xc)
Ec (= Yc)
X-count
Percentage of X-count to ICE2
Y-count
Percentage of Y-count to ICE2
ICE
ICE2
Percentages of ICE2 to ICE2, that is, 100%

7. Build the second table 4.6.2B with rows corresponding to each pair of values of Xc and Yc with the following information:
Ic (= Xc)
Ec (= Yc)
Pure-internal-c-ticks-count
Percentage of the above value to ICE
Pure-external-c-ticks-count

Percentage of the above value to ICE
Composite-c-pairs-count
Percentage of the above value to ICE
ICE
Percentage of ICE to ICE, that is, 100%
ICE2

** Table 4.6.2A: Simulated c-hulchul values (raw data)

Ic	Ec	X-Count	Y-Count	ICE	ICE2
		41,002	1,999,986	2,000,000	2,040,988
2	98	2.0089%	97.9911%		100%
		222,983	1,997,561	2,000,000	2,220,544
10	90	10.0481%	89.9582%		100%
		494,167	1,976,357	2,000,000	2,470,524
20	80	20.0025%	79.9975%		100%
		816,455	1,907,244	2,000,000	2,723,699
30	70	29.9760%	70.0240%		100%
		1,168,109	1,753,863	2,000,000	2,921,972
40	60	39.9767%	60.0233%		100%
		1,500,016	1,501,736	2,000,000	3,001,752
50	50	49.9714%	50.9714%		100%

** Table 4.6.2B: Simulated hulchul values (computed values)

C1	C2	C3	C4	C5	C6	C7
Ic	Ec	Min I	Min E	Max C	ICE	ICE2
		15	1,958,998	40,987	2,000,000	2,040,988
2	98	0.0008%	97.9499%	2.0494%	100%	
		2,440	1,777,017	220,543	2,000,000	2,220,544
10	90	0.1220%	88.8509%	11.0272%	100%	
		23,644	1,505,833	470,523	2,000,000	2,470,524
20	80	1.1822%	75.2917%	23.5262%	100%	
		92,757	1,183,545	723,698	2,000,000	2,723,699
30	70	4.6379%	59.1773%	36.1849%	100%	

C1	C2	C3	C4	C5	C6	C7
Ic	Ec	Min I	Min E	Max C	ICE	ICE2
		246,138	831,891	921,971	2,000,000	2,921,972
40	60	12.3069%	41.5946%	46.0986%	100%	
		498,265	499,984	1,001,751	2,000,000	3,001,752
50	50	24.9133%	24.9992%	50.0876%	100%	

4.6.3 Equivalence of Probabilistic and Simulation Methods

(We are presenting Table 4.6.1B and Table 4.6.2B on the same page.)

** Table 4.6.1B: Probabilistic hulchul values with ICE = 100

C1	C2	C3	C4	C5	C6	C7
Ic	Ec	Min. I	Min. E	Max. C	ICE	ICE2
					I+C+E	Ic + Ec
2.0408	99.9992	0.0008	97.9592	2.0400	100	102.0400
2%	98%					
11.0976	99.8780	0.1220	88.9024	10.9756	100	110.9756
10%	90%					
24.7059	92.8235	1.1765	75.2941	23.5294	100	123.5924
20%	80%					
40.8621	95.3448	4.6551	59.1379	36.2069	100	136.2069
30%	70%					
58.4615	87.6923	12.3077	41.5385	46.1538	100	146.1538
40%	60%					
75.0000	75.0000	25.0000	25.0000	50.0000	100	150.000
50%	50%					

** Table 4.6.2B: Simulated hulchul values (computed values)

C1	C2	C3	C4	C5	C6	C7
Ic	Ec	Min I	Min E	Max C	ICE	ICE2
		15	1,958,998	40,987	2,000,000	2,040,988
2	98	0.0008%	97.9499%	2.0494%	100%	
		2,440	1,777,017	220,543	2,000,000	2,220,544
10	90	0.1220%	88.8509%	11.0272%	100%	

C1	C2	C3	C4	C5	C6	C7
Ic	Ec	Min I	Min E	Max C	ICE	ICE2
		23,644	1,505,833	470,523	2,000,000	2,470,524
20	80	1.1822%	75.2917%	23.5262%	100%	
		92,757	1,183,545	723,698	2,000,000	2,723,699
30	70	4.6379%	59.1773%	36.1849%	100%	
		246,138	831,891	921,971	2,000,000	2,921,972
40	60	12.3069%	41.5946%	46.0986%	100%	
		498,265	499,984	1,001,751	2,000,000	3,001,752
50	50	24.9133%	24.9992%	50.0876%	100%	

Comparison of Tables 4.6.1B and 4.6.2B:

1. In column C1 of 4.6.1B, the percentages in the first pair are percentages to ICE2 and the percentages in the second pair of are percentages to ICE. In column C1 of 4.6.2B, the percentages are to ICE.
2. The corresponding values in columns C3, C4 and C5 of the two tables are almost equal.
3. The probabilistic method provides limits on the values I, E and C in an analytical form. (See inequalities 4.6A, 4.6B and 4.6C)
4. The simulation method **does not** provide limits on the values I, E and C in an analytical form but the method is more assured.
5. The simulation method Table 4.6.2B shows both counts and the percentages in the columns C3, C4 and C5 whereas the probabilistic method Table 4.6.1B shows only the percentages.
6. ICE2 = ICE + C in each row of both tables as expected.

Note: For simulation method, actual generation of tables 4.6.2A and 4.6.2B took less than two minutes on a laptop with modest capability.

4.7 Proposition #2

Both internal and external c-hulchul of a common body must be greater than zero percent and less than 100 percent of the total c-hulchul of the body for all common bodies in all sufficiently large h-domains.

Is there any physical evidence that a common body cannot attain 100% internal c-hulchul in a sufficiently large h-domain? Possible examples are:

4.7.1 A Particle Accelerator a Possible Example

A particle accelerator is used to make particles, inside it, move at speed as close to the speed of light as possible. In terms of c-hulchul, this is an attempt to bring internal c-hulchul of the particle accelerator as close as possible to the universal c-hulchul CH(U), that is, 100%. In this attempt, there is a catastrophic change in the state of particles participating in the particle accelerator; in fact, that is the goal of the scientists. In terms of c-hulchul, we will interpret this phenomenon as if the attempt to achieve maximum internal c-hulchul of 100% aborts of its own accord.

4.7.2 An Atom another Possible Example

On the assumption that electrons do move in an atom: Could an atom, with a large number of electrons, attain 100% internal c-hulchul **unless** there were some kind of compartmentalization of electrons so the electrons in a compartment move in sync, that is, as one body? Are the shells and the sub-shells in an atom the possible compartments in the atom to control internal c-hulchul (of the atom)?

4.8 Proposition #3

Light and photons do not inherit external hulchul of their outer bodies nor do they contribute to the internal hulchul of their outer bodies.

We believe it is because of the above proposition that the speed of light is independent of the speed of its source. Light and photons are not constituent parts of any body other than the U body; a possible exception may be brief periods of transitions due to interaction between photons and matter. Light and photons keep on moving wavy straight, unless obstructed by other bodies, without inheriting motion of other bodies.

Note: We believe Proposition #3 is certainly true in a vacuum; it may be necessary to adjust it for motion of phantom bodies in media such as air and water.

4.9 Time Dilation

We will analyze **relativistic time dilation** in the context of hulchul. We will call the time dilation in terms of hulchul as **hulchulic time dilation**. There are some critical differences between relativistic time dilation and hulchulic time dilation:

- Distance traveled by a body b versus number of external c-ticks made by b.
- Assumed relative motion versus absolute motion.
- Classical time versus u-time and bs-time.
- Hulchulic time dilation of a body b implies using the absolute motion of b.
- **Even relativistic time dilation of a body b implies using the absolute motion of b contrary to the assumption that it is based on relative motion of b.**

Consider Figure 4.9 below:

- b, b3 and b4 are ECH-equivalent bodies since b3 and b4 have no motion in b. Motion of b3 and b4 is all inherited from the body b. b3 and b4 have time dilation even if they have no motion relative to b; in fact, b, b3 and b4 have the same time dilation.
- b1 and b2 have their own external motion in b aside external motion inherited from b. b1, b2 and b3 have different external motions and, accordingly, different dilations. Similarly, b1, b2 and b4 have different external motions and, accordingly, different time dilations.

** **Figure 4.9**: Time dilation using absolute motion

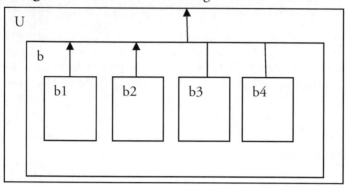

4.9.1 Relativistic Time Dilation

** **Figure 4.9.1**

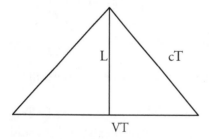

T = Time taken by the light when the Earth is moving.
V = Speed of the Earth.
c = Speed of the light.
T_{REST} = Time taken by the light if the Earth is at rest.
L = Distance traveled by the light one way when Earth is at rest.

$L = cT_{REST}$
$(cT)^2 = L^2 + (VT)^2 = (cT_{REST})^2 + (VT)^2$
$T^2(c^2 - V^2) = (cT_{REST})^2$
Since $T = T_{REST}$ as per **Michelson Morley's experiment**, therefore

Relativistic Time Dilation: $\dfrac{T}{T_{REST}} = \dfrac{1}{\sqrt{1 - \dfrac{v^2}{c^2}}}$ (4.9.1)

4.9.2 Hulchulic Time Dilation

Interpreting c-hulchuls in terms of classical motion and time: Here, instead of using distance traveled by a body b, we use number of c-ticks made by b.

ICE = uTu or u = ICE / Tu = CH(U) / Tu
CH(U) = ICE if Proposition #2 is true.
Ec = vTu or v = Ec / Tu
I = uTb
v/u = Ec / ICE = CH[b] / CH(U)

Where
   u is the number of universal c-ticks per unit of universal time.
   Tu is the number of the above units of universal time for CH(U).
   v is the number of external c-ticks per unit of universal time.
   Tb is the number of units of universal time for CH((b)).
   Using (4.9.1) and replacing V / c by v / u or CH[b] / CH(U)

Hulchulic time dilation = $\dfrac{1}{\sqrt{1 - \dfrac{(CH[b])^2}{(CH(U))^2}}}$ (4.9.2A)

$= \dfrac{CH(U)}{\sqrt{(CH(U))^2 - (CH[b])^2}}$ (4.9.2B)

$$= \frac{1}{\sqrt{1 - \dfrac{v^2}{u^2}}} \qquad\qquad \textbf{(4.9.2C)}$$

Time dilation depends on the absolute motion of a body: All three hulchulic time dilation expressions (4.9.2A), (4.9.2B) and (4.9.2C) depend on the absolute motion of a body and not on relative motion of the body. The same must be true of relativistic time dilation.

Question: What is the significance of $SQRT((CH(U))^2 - (CH[b])^2)$ in (4.9.2B)? What does it represent physically?

Possibly, some kind of bs-time of b.

Question: What is **bs-time of a body b? ch(b), or ch((b)) or** $SQRT((CH(U))^2 - (CH[b])^2)$**?**
This question arises because: As per (4.9.2A), hulchulic time dilation depends on external c-hulchul Ec only. (CH(U) is a constant.) If Ec is known, then I, and not Ic, can be determined uniquely, since I+Ec = CH(U) (subject to the veracity of Proposition #1). We may also say: hulchulic time dilation depends on pure internal c-hulchul I only. There are now three reasons why I, and not Ic, may be the bs-time of a body b:

1. If I is the bs-time, then Proposition #1 must be true—another reason for the veracity of Proposition #1.
2. As far as universal time (u-time) is concerned, it does not matter whether I is the bs-time or Ic is the bs-time, since ch(U) = ch((U)). The same is true for elementary bodies, since ch(e) and ch((e)) are null for any elementary body e.
3. Ic = I+C. Since C is common to both Ec and Ic, the C part of Ic may be ignored for this purpose.
4. Another possibility is that $SQRT((CH(U))^2 - (CH[b])^2)$ has a physical significance here.

We are unsure.

4.9.3 Comparison of Relativistic and Hulchulic Time Dilations

** **Table: 3.4.3: Correspondence between Relativistic and Hulchulic Time Dilation Terms**

Time Outage for body b		Time Dilation for the Earth	
u	Universal c-ticks per unit of u-time.	c	Speed of light
v	External c-ticks per unit of u-time.	V	Speed of the Earth
Tb	Units of bs-time of b.	T_{REST}	Time used by light if the Earth is at rest.
Tu	Units of u-time.	T	Time used by light if the Earth is not at rest.
		$L = cT_{REST}$	Distance traveled by light if the Earth is at rest.
ch(b)	uTb Number of internal c-ticks of b.	$T_{REST} = L / c$	Time used by light if Earth is at rest.
ch(U)	uTu Number of universal c-ticks.	cT	Distance traveled by light if the Earth is not at rest.
ch[b]	vTu Number of external c-ticks of b.	VT	Distance traveled by the Earth.

4.9.4 Time Dilation for ECH-equivalent Bodies

Time outage for two ECH-equivalent lineal bodies b1 and b may be different but their hulchulic time dilation must be the same, since ECH-equivalent bodies have the same external c-hulchul. This is also implied by (4.9.2), since the expression in (4.9.2A) involves only CH(U) and CH[b], which have the same value for two ECH-equivalent lineal bodies.

Example: b1 and b are ECH-equivalent lineal bodies; b1 ⊂ b

(This is not empirical data; this is just to illustrate the concepts.)

	Change from b to b1	b	b1
Ec	No change	75%	75%
I	No change	25%	25%
C	Decreases by ΔC	30%	27% (ΔC = 3%)
E	Increases by ΔC	45%	48%
Ic	Decreases by ΔC	55%	52%
ICE	No change	100%	100%
	Hulchulic Time dilation using (4.9.2A)	1.51	1.51
E	Time outage	45%	48%

Note:

Ec = E + C

Ic = I + C

ICE = I + C + E = I + Ec = Ic + E

CHAPTER 5

(Conclusions And Summary)

Contents

5.1 Conclusions

5.1.1 Areas of Unfinished Research Work

More research needs to be done in the following areas:

1. Rotational motion.
2. Accelerated and retarded motions.
3. Multi-directional motion:
   Explosive motion
   Implosive motion
   Waves
4. Veracity of **Axiom #2 of elementary bodies and Proposition #1**.
   (Either of the two implies the other.)
5. How to determine **concurrency** of ticks logically?
6. What exactly is the difference between **time outage and time dilation**?
7. Time outage—precise mathematical expression of time outage.

8. **ECH-equivalent lineal bodies** must have the same time dilation. This physical fact should be exploited.
9. **Middle bodies (having Ic = Ec).**

5.1.2 Bio-psychological Systems and Hulchul

- **A Focused Mind and Its Hulchul**: In general, we do better when we are focused. When we are focused, our mental productivity in solving problems, in particular, may increase many-fold. Bio-psychological systems also have internal and external hulchuls, which change dynamically. Can this change be associated with different mental states? More specifically, is there a relationship between external hulchul and a focused mental state? In the state of a focused mind, it appears that the external hulchul of our bio-psychological system increases and as a result the internal hulchul decreases.

- **Different States of Mind and Their Hulchul**s: In a state of anxiety, fear, pain, disturbed sleep, boredom and vacillation and confusion, internal hulchul appears to increase and external hulchul appears to decrease. Whereas, in a state of relaxation, good feeling, fun, sound sleep, determination and positive attitude, external hulchul appears to increase.

The preceding two statements are not based on any empirical data.

5.1.3 A Hypothetical Experiment to Detect Absolute Motion of a Body in the Universe

Whether a train is moving or it is stationary, an electric light bulb falling from the ceiling of the train will hit the floor of the train at the same point X exactly below the holder of the bulb. This is by the principle of relativity. In terms of hulchul, and this is probably because of the light bulb inheriting the motion of its outer bodies including the train.

Now consider a single photon shooting from the same light bulb. Will it hit point X or not? Some of the possibilities are:

1. The photon hits point X.
2. The photon may not hit point X since a photon does not inherit the motion of its outer bodies and as a result the Principle of Relativity (for motion) does not apply to a photon; instead, it will hit a point that depends on the resultant speed and direction of:
    A. The train.
    B. The train and the Earth.
    C. The absolute motion of the train in the universe taking into account the motion of all outer bodies of the train.

We can also use here a **hypothetical device,** which can shoot photons, one at a time, to a screen exactly at the same spot. The device has been presumably tested in places where photons shot by the device are not interfered with by the matter around, if any, to make it possible for the photons to be shot to the same spot on the screen. We assume it will work on the Earth, too, if we have taken enough precautions.

Now we test the device on the Earth. Since the Earth has rotational motion and orbital motion and also it inherits the motion of the Solar System, Milky Way and possibly further outer bodies, we should take into account the total of all of these motions, that is, the absolute motion of the Earth. The screen and the device will move along with the absolute motion of the Earth but the absolute motion of the Earth will not have any effect on the travel of a photon from the device to the screen as the motion of light and photons is not affected by the motions of their outer bodies. The principle of relativity, that laws of physics are the same in all uniformly moving frames of references, does not apply to light and photons. Because of the exception for light and photons and because of the (absolute) motion of the Earth, the photons should not hit the same spot on the screen. Since the travel time of a photon will be extremely small, the photons should not hit very far from each other on the screen. The pattern where the photons hit the screens should tell us about the nature of the absolute motion of the Earth.

Note: Photons may be disturbed by the presence of matter around them but the motion of their outer bodies does not add to the motion of photons.

5.2 Summary

Goal: To explore for an intrinsic, deeper relation between motion and time in the light of time dilation.

Bodies and Their Motion

- **Inner Body and Outer Body Relationship:** If a body $b1$ is a constituent part of a body b, then $b1$ is an inner body of b, and b is an outer body of $b1$. We denote this relationship by $b1 \subset b$.
- **Internal, External and Composite Motion of a Body:** If $b1$ moves in b, then $b1$ has **external motion** and b has **internal motion**. The two motions are two views of the same motion.
  If internal motion and external motion of a body b are concurrent then b has composite motion.
- **Outer Body as a Frame of Reference**. For the motion of a body b, we will use mostly an outer body B of b as a **frame of reference**.
- **Extended and Absolute Motion of a Body**: Suppose $b \subset B \subset B1$, b moves in B, and B moves in $B1$, then b moves in $B1$ and this will be called an **extended external motion** of b in $B1$. Extended external motion of b in U will be called an **absolute external motion** of b. (Internal motion of a body is always absolute).
- **Inheritance of Motion of a Body:** A body b goes wherever its outer body B goes. This is called: Inner body b inherits external motion of B.
- **The Four Types of Bodies**:
  The U body—the universe as a whole
  Elementary Body—has no inner body
  Phantom Body—neither inherits motion of other bodies nor do other bodies inherit motion of phantom bodies
  Common Body—different from the bodies of the above three types

- **Lineal and Non-Lineal Bodies**: Two bodies are lineal, if exactly one of the two is an inner body of the other, otherwise the two bodies are non-lineal.

Ticks, Prime Ticks and Chained Ticks

- **Ticks of Motion of a Body**: Motion of a body b in body B is perceived as a repetition of the following two steps together:
  1. b moves in B.
  2. b pauses and/or it changes the direction of its motion.
  We call the motion during a single move a **tick** of the body b in B, denote it by **t[b, B],** and say: **b ticks in B**.
- **Alternative Definition of Ticks**: A tick of a body is the movement of the body during an instant of time.
- **Instant of Motion**: An instant of motion of a body is the movement of the body during an instant of time.
- **Prime Tick (P-Tick) and Chained Ticks**: Suppose a body b ticks in body B. Let t1 = t[b, B]. If there exists a body b0 such that b ⊂ b0 ⊂ B, t[b, b0] and t[b0, B], then t1 is called a **chained tick of b in B.** If there exists no b0, as in the preceding manner, then t1 is called a **prime tick of b in B.**
- **Significance of Prime Ticks:**
  o   A prime tick is the smallest possible movement of a body.
  o   Prime ticks are to motion what elementary particles are to matter and what photons are to light. Prime ticks are all-pervasive in the universe.
  o   **Wavy straight Motion of Fast Moving Bodies:**
      A fast moving body, such as an elementary particle, has wavy straight motion rather than straight motion.
  o   **Difference between a Slow Motion and a Fast Motion:** A Fast motion involves more p-ticks than a slow motion does.
- **Empty, Non-empty Ticks**: If b ⊂ B, and b does not tick in B, then t[b, B] is called an **empty tick**, else t[b, B] is called a **non-empty** tick.
- **Internal and External Ticks**: If t1 = t[b, B], then t1 is called an **external tick** of b and an **internal tick** of B.

- **Object and Reference Bodies of a Tick**: In a tick t[b, B], b is called the **object body** of the tick and B is called the **reference body** of the tick.
- **Property of Ticks:** A non-empty tick can be decomposed into all prime ticks.
- **Length of a tick**: The number of prime ticks in a tick is called the length of the tick. A tick is an empty tick, or a prime tick or a chained tick according as the length is 0, 1 or > 1.
- **Internal, External and Composite ticks of a body**.
  Tick of b in B: t[b, B]; b ticks in B.
  External tick of b: t[b]; b ticks (in some body).
  Internal tick of b: t(b); some body ticks in b.
  Pure external tick of b: t[[b]]; b has an external tick but no concurrent internal tick.
  Pure internal tick of b: t((b)); b has an internal tick but no concurrent external tick.
  Composite tick of b: t[(b)]; b has concurrent internal and external ticks.
- **Extended Prime Tick (XP-Tick)**: A prime tick of an outer-body of a body b is called an **external xp-tick of b**. A prime tick of an inner body of a body b is called an **internal xp-tick** of b.
- **Property of XP-Ticks**: An internal xp-tick of a body b is either an internal prime tick of b or it is concurrent with an internal prime tick of b. An external xp-tick of a body b is either an external prime tick of b or it is concurrent with an external prime tick of b.
- **Maximal Tick**: An external tick t of a body b in B is called a **maximal external tick** of b if B has no external tick concurrently with t. An external tick t of a body b in B is called a **maximal internal tick** of B if b has no internal tick concurrently with t.
- **Inheritance of Ticks:** Suppose b1 ⊂ b. Body b1 inherits all external ticks of b as external ticks. Body b inherits all internal ticks of b1 as internal ticks.
- **Axiom of Exclusion**: Same tick cannot be both an internal tick and an external tick of a body. Further: suppose b1 ⊂ b; body b1 cannot inherit an external tick of b as an internal tick. Body b cannot inherit an internal tick of b1 as an external tick.

- **Principle of Relativity:** It should be possible to derive the Principle of Relativity (motion part of it) from the inheritance of motion (external c-hulchul).

- **Strong Transitivity of Ticks:** Suppose b1 ⊂ b ⊂ B. If t[b1, b] **or** t[b, B], then t[b1, B] and vice versa.
  If a passenger walks in train **or** the train moves, the passenger moves relative to the Earth and vice versa.

- **Independent Ticks:** Two ticks are said to be independent if no prime tick in either tick is inherited from a prime tick of the other tick.

- **Native Tick:** A native tick is a prime tick not inherited from any other prime tick.

  o **Difference between Classical and Hulchulic Motion:** Classical motion of a body b is generally based on native ticks of a body b. Hulchulic motion is based on both native and inherited ticks.

  o **Difference between Motion of Photons and Light and Motion of Elementary Particles:** Photons and light have only native ticks, whereas an elementary particle may have both native and inherited prime ticks.

- **Equation of a Tick:**
  t[b1] = t[b1, b] ∪ t[b]
  That is, if b1 ⊂ b and b1 ticks, then either b1 ticks in b and/or b ticks.

- **Axiom of Concurrent External Ticks:** A body cannot have two concurrent external ticks unless their reference bodies are lineal. (There is no such restriction on internal ticks of a body. See Figures: 1.5.24F, G and H.)

- **Sharing of an Electron by Two Atoms** is neither concurrent nor continuous; it alternates between the two atoms as per the axiom of concurrent external ticks. (See Figure 1.5.24H.)

Extended Prime Hulchuls (Sets of XP-Ticks)

- **External Extended Prime Hulchul of a Body xph[b]:** xph[b] is the set of all independent p-ticks of a body b or outer bodies of b.

- **Internal Extended Prime Hulchul of a Body xph(b)**: xph(b) is the set of all independent internal p-ticks of a body b or inner bodies of b.
- **Types of Extended Prime Hulchuls (xp-hulchuls)**:
  o Internal xp-hulchul: xph(b)
  o External xp-hulchul: xph[b]
  o Relative xp-hulchul of b1 in b: xph[b1, b]
  o Composite internal xp-hulchul: xph[I(b)] =
    {x | x ∈ xph(b) and x is concurrent with some y ∈ xph[b]}
  o Composite external xp-hulchul: xph[E(b)] =
    {x | x ∈ xph[b] and x is concurrent with some y ∈ xph(b)}
  o Pure internal xp-hulchul: xph((b)) =
    {x | x ∈ xph(b) and x is not concurrent with any y ∈ xph[b]}
  o Pure external xp-hulchul: xph[[b]] =
    {x | x ∈ xph[b] and x is not concurrent with any y ∈ xph(b)}
  o Total xp-hulchul: xph{b} = xph(b) ∪ xph[b]
    Note: xph(b) and xph[b] are disjoint due to axiom of exclusion.

Theory of Concurrency of Events with Focus on Ticks as Events

- **E-Set**: A set of events is called an **e-set**.
- **Ticks are events;** xp-hulchuls are e-sets.
- **Concurrency-Event (C-Event)**: A class of events is called a **c-event** if all members in the class are concurrent.
- **Concurrency Set (C-Set)**: An e-set is called a **concurrency set(c-set)** if no two members of the e-set are concurrent.
- **Concurrency Disjoint (C-Disjoint) E-Sets**: Two e-sets are said to be **concurrency disjoint (c-disjoint)** if no member from one e-set is concurrent with any member of the other e-set.
- **Concurrency Operator CNCY**: Suppose P is an e-set. Then **CNCY(P)** = The set of all concurrency events derived from the events in P. CNCY(P) is a c-set.
  Examples: ch(b) = CNCY(xph(b)), ch[b] = CNCY(xph[b]).
  Property: If P ⊂ Q, then CNCY(P) ⊆ CNCY(Q).
- **Concurrency Difference Operator CDIF**:
  CDIF(P, Q) = {x | x ∈ P and x is not concurrent with any member of Q}.

In general, CDIF(P, Q) and CDIF(Q, P) are different.
CNCY(CDIF(P, Q)) = CDIF(CNCY(P), CNCY(Q)).
CDIF(P, Q) ⊆ P; CDIF(Q, P) ⊆ Q.

- **Concurrency Intersection Operator CINT:**
  CINT(P, Q) = {x | x ∈ P and x is concurrent with some member of Q}.
  In general, CINT(P, Q) and CINT(Q, P) are different.
  CNCY(CINT(P, Q)) = CINT(CNCY(P), CNCY(Q)).
    CINT(P, Q) ⊆ P; CINT(Q, P) ⊆ Q.

Concurrency Hulchuls (C-Hulchuls)

- **Types of Concurrency Hulchuls (C-hulchuls):**
  o Internal c-hulchul: ch(b) = CNCY(xph(b))
  o External c-hulchul: ch[b] = CNCY(xph[b])
  o Relative c-hulchul of b1 in b: ch[b1, b]
  o Composite c-hulchul: ch[(b)] = ch(b) ∩ ch[b]
     = CNCY(xph(I[b]) = CNCY(xph[(E(b)]
     = CNCY(CINT(xph(b), xph[b]))
     = CNCY(CINT(xph[b], xph(b)))
  o Pure internal c-hulchul: ch((b)) = ch(b) − ch[(b)]
  o = ch(b) − ch[b] = CNCY(CDIF(xph(b), xph[b]))
  o Pure external c-hulchul: ch[[b]] = ch[b] − ch[(b)]
  o = ch[b] − ch(b) = CNCY(CDIF(xph[b], xph(b)))
  o Total c-hulchul:
     ch{b} = ch(b) ∪ ch[b] = ch((b)) ∪ ch[(b)] ∪ ch[[b]]
     Note: ch((b)), ch[[b]] and ch[(b)] are disjoint.
     ch{U] = ch(U) and ch{e} = ch[e], where e is an elementary body.

- **Notations for Counts of Hulchuls: (by examples)**
  XPH(b) = count(xph(b))
  CH(b) = count(ch(b))

If the underlying body b is known:
I = CH((b)) = count(ch((b)))
E = CH[b] = count(ch[[b]])
C = CH[(b)] = count(ch[(b)])

Ic = CH(b) = count(ch(b)) = I + C
Ec = CH[b] = count(ch[b]) = E+ C
ICE = I + C + E = CH{b} = count(ch{b})

- **Equation of Relative XP-Hulchul**:
  xph[b1] = xph[b1, b] ∪ xph[b]
- **Inheritance Properties of XP-Hulchuls**:
  o   Internal xp-hulchuls xph(b) and xph((b)) and their counts increase or remain the same outer-body-ward.
  o   External xp-hulchuls xph[b] and xph[[b]] and their counts increase or remain the same inner-body-ward.
  o   Composite xp-hulchul, in general, has no inheritance property.
  o   Total xp-hulchul xph{b} and its count increase or remain the same outer-body-ward.

- **Equation of Relative C-Hulchul**:
  ch[b1] = ch[b1, b] ∪ ch[b], where b1 ⊂ b
- **Inheritance Properties of C-Hulchuls**:
  o   Internal c-hulchuls ch(b) and ch((b)) and their counts increase or remain the same outer-body-ward.
  o   External c-hulchuls ch[b] and ch[[b]] and their counts increase or remain the same inner-body-ward.
  o   Composite c-hulchul, in general, has no inheritance property.
  o   Total c-hulchul ch{b} and its count increase or remain the same outer-body-ward. (See Proposition #1.)
- **Middle Bodies, Middle C-hulchuls**: Since Ic increases outer-body-ward and Ec increases inner-body-ward, there may exist a body having Ic = Ec. Such a body is called a **middle body** and its c-hulchul is called a **middle c-hulchul**.

- **Lineal Range of Two Bodies**: Relative c-hulchul of a body b in B depends on the two bodies, where b ⊂ B. We call a pair of two lineal bodies b and B a **lineal range** of the two bodies. Wider the range of the two bodies b and B, larger the value of ch[b, B]. For example: Suppose b2 ⊂ b1 ⊂ b ⊂ B ⊂ B1. Then ch[b1, b] ⊆ ch[b1, B]; ch[b1, b] ⊆ ch[b2, b]; ch[b1, b] ⊆ ch[b2, B].

- **CH-Equivalent Bodies:** Two body having the same corresponding c-hulchul for each type.
- **ECH-Equivalent Bodies:** Two bodies are said to be ECH-equivalent if they are not CH-equivalent and external c-hulchuls of the two bodies are the same.
  Example: The Earth and a stationary car, on the surface of the Earth, are ECH-equivalent.
- **Universal xp-hulchul**: xph(U) is called the universal xp-hulchul. xph(U) has all prime ticks in the universe. Possibly, xph(U) has all events in the universe if we assume each event involves some kind of motion.
- **Universal c-hulchul**: ch(U) is called the universal c-hulchul. C-ticks of U are called universal c-ticks. We assume universal c-ticks represent all possible instants of time in the universe.
- **Universal Time (u-time):** ch(U) represents **universal time (u-time).**
- **Body-Specific Time (bs-time):** In line with universal time, ch(b) represents **body-specific time (bs-time)** for the body b.
- **Hulchul Domains (h-domain)**: A subsequence of ch(U) is called an h-domain. An h-domain may be continuous or discontinuous. An h-domain is like a time interval with the difference: whereas, time interval is continuous, an h-domain may be continuous or discontinuous.
- **Axiom 1 of Elementary Bodies**: Each common body has at least one elementary body as an inner body.
- **Axiom 2 of Elementary Bodies**: An elementary body e has an xp-tick during each universal c-tick. The xp-tick may be a native tick or an inherited tick of e. As a result: **ch[e] = ch(U).**
- **Proposition #1**: Total c-hulchuls of any two bodies is the same no matter how small or large the two bodies are.
  ch{b} = ch{q} = ch{U} = ch(U) for any bodies b and q.
  o   Proposition #1 is true if and only if the Axiom 2 of elementary bodies is true. However, ch{b1} $\subseteq$ ch {b} unconditionally for b1 $\subset$ b.
- **Convergence of the Three Concepts: Internal C-hulchul of the U body, Universal Time and Motion of Light**: It appears that the three concepts converge into one and the same concept. (Note that ch[light] = ch(U).)

Estimation of C—The Number of Composite C-ticks

- **Estimates of I and E**: If there are three consecutive c-ticks of the same type (internal or external), then the middle c-tick must be a pure c-tick of the same type.
- **Composite C-pair**: It is a pair of consecutive internal and external c-tick. A composite c-tick must be a c-pair but a c-pair may or may not be a composite c-tick. We can estimate the number of C-pairs easily. In fact, R.H.S of (4.6A) is this number.
- **Estimate of C**: There is no theoretical way to know, at this stage of this research, whether an internal tick and an external tick are concurrent or not. We can, at best, find a maximum limit on the value of C by using estimated number of c-pairs.

- **Probabilistic Limits on the Values of I, E and C in case of a Common Body:**

$$\textbf{\textit{Maximum value of C}} \leq \frac{IcEc(Ic + Ec)}{(Ic)^2 + (Ec)^2 + IcEc} \qquad (4.6A)$$

Since I = Ic – C and E = Ec – C, therefore:

$$\textbf{\textit{Minimum value of I}} \geq \frac{(Ic)^3}{(Ic)^2 + (Ec)^2 + IcEc} \qquad (4.6B)$$

$$\textbf{\textit{Minimum value of E}} \geq \frac{(Ec)^3}{(Ic)^2 + (Ec)^2 + IcEc} \qquad (4.6C)$$

This means for a common body:
   0% < I ≤ Ic < 100%
   0% < E ≤ Ec < 100%
   C > 0;

- **Proposition #2:** Both internal and external c-hulchuls of a common body must be greater than 0% and less than 100% of the total c-hulchul of a common in all sufficiently large h-domains.
- **Some Possible Consequences of the Proposition #2**

- o **Particle Accelerator:** Proposition #2 may be the reason behind the working of a particle accelerator.
- o **Internal Structure of an Atom:** Compartmentalization of the electrons of an atom, with a large number of electrons, may be a reason to keep internal c-hulchul of an atom below 100%.
- **Time Outage of an External C-tick A body**: An external c-tick x of a body b during which b does not experience any body-specific time.
- **Time Outage of a body** b is the pure external hulchul ch[[b]]. Time outage is the cause of **time dilation**.
- **Time outages of two ECH-equivalent lineal bodies, in general. are different.**
- **Time Dilation of a body is defined as:**

$$\text{Hulchulic Time dilation} = \frac{1}{\sqrt{1 - \frac{(Ec)^2}{(CH(U))^2}}} \qquad (3.14.1B)$$

Hulchulic time dilation is based on the absolute motion of the body.

Time dilations of two ECH-equivalent lineal bodies are the same.

- **Proposition #3**: Light and photons do not inherit external c-hulchul of their outer bodies nor do they contribute to the internal c-hulchul of their outer bodies.
- **A Hypothetical Equation of C-Hulchul of a Body b:**

$$(I + C)^2 + (E + C)^2 = (I + C + E)^2 \ (4.4B)$$

The above intuitional equation leads to the following expression in terms of hulchulic motion, which is very similar to the time dilation factor:

$$Or \quad \frac{Tu}{Tb} = \frac{1}{\sqrt{1 - \frac{v^2}{u^2}}} \qquad (4.4D)$$

Where
   u = the number universal c-ticks per unit of time

Tu = the number of units of u-time

$I + C + E = u^*Tu = ch(U)$

v = rate of external c-ticks

$E + C = v^*Tu$

Tb = number of units of bs-time

$(I + C) = u^*Tb$

However, (4.4B) is untenable if Proposition #1 is true. This is because, we have proven that we cannot have more than one independent equation in I, C and E due to existence of ECH-equivalent bodies. For example, the Earth and a stationary car on the surface of Earth have the same external c-hulchul but the corresponding values of Ic are different. **This means Tu/Tb on the L.H.S of (4.4D) does not represent time dilation but the R.H.S of (4.2D) does. Therefore, CH(b) does not does not lead to time dilation of a body b.**

LISTS OF NOTATION AND FIGURES

List of Hulchul Related Notations

** Table of Notations

Type	Notation	Description
Inner/ outer body relationship	**b ⊂ B**	b is an inner body of B; B is an outer body of b.
	b ⊆ B	b ⊂ B or b = B.
Tick	**t1 = [b, B]**	b ticks in B; t1 is an external tick of b in B; t1 is an external tick of b; t1 is an internal tick of B.
	t1 =[b, B:x]	Tick t occurs during a universal c-tick x (an instant of time).
	t1= [b, B;n]	Length of the tick t1 is n. The tick is empty, a prime tick or a chained tick according as n is 0, 1 or > 1.
Single body based ticks	**t[b]**	An external tick of b
	t(b)	An internal tick of b.
	t[(b)]	A composite tick of b.
	t[[b]]	A pure external tick of b.
	t((b))	A pure internal tick of b.
Inheritance of ticks	**t <= T**	Tick t inherits tick T. Tick T is inherited by tick t.
	T => t	Same as t <= T. (T implies t.)
Order of events	**E1 → E2**	Event E1 occurs after event E2.
	E1 ← E2	Event E1 occurs before event E2.
Concurrency of events	**E1 ↔ E2**	Events E1 and E2 are concurrent.

Type	Notation	Description
Prime hulchul (p-hulchul)	**ph(b)**	The internal prime hulchul of b.
	ph[b]	The external prime hulchul of b.
	ph[E(b)]	The composite external prime hulchul of b.
	ph[I(b)]	The composite internal prime hulchul of b.
	ph[[b]]	The pure external prime hulchul of b.
	ph((b))	The pure internal prime hulchul of b.
	ph[b, B]	Relative prime hulchul of b in B.
	ph{b}	The total prime hulchul of b.
Extended Prime hulchul (xp-hulchul)	**xph(b)**	The internal xp-hulchul of b.
	xph[b]	The external xp-hulchul of b.
	xph[E(b)]	The composite external xp-hulchul of b.
	xph[I(b)]	The composite internal xp-hulchul of b.
	xph[[b]]	The pure external xp-hulchul of b.
	xph((b))	The pure internal xp-hulchul of b.
	xph[b, B]	Relative external xp-hulchul of b in B.
	Xph{b}	The total hulchul xp-hulchul of b.
Concurrency hulchul (c-hulchul)	**ch[b]**	The external c-hulchul of b.
	ch(b)	The internal c-hulchul of b.
	ch[(b)]	The composite c-hulchul of b.
	ch[[b]]	The pure external c-hulchul of b.
	ch((b))	The pure internal c-hulchul of b.
	ch[b, B]	Relative c-hulchul of b in B.
	Ch{b}	The total c-hulchul of b.
Operator	**CNCY**	CNCY(P) = Concurrency events of P where P is a set of events.
	CDIF	CDIF(P, Q) = Concurrency difference of P and Q where P and Q are sets of events.
	CINT	CINT(P, Q) = Concurrency intersection of P and Q where P and Q are sets of events.
Count	**count**	count(S) = count of members in the set S.
	PH[b]	PH[b] = count(ph[b]) Similar notations for other p-hulchuls.

Type	Notation	Description
	XPH[b]	XPH[b] = count(xph[b])
		Similar notations for other xp-hulchuls.
	CH[b]	CH[b] = count(ch[b])
		Similar notations for other c-hulchuls.
	Ec	Ec = CH[b], underlying body b is assumed.
	Ic	Ic = CH(b), underlying body b is assumed.
	E	I = CH[[b]], underlying body b is assumed.
	I	I = CH((b)), underlying body b is assumed, underlying body b is assumed.
	C	C = CH[(b)], underlying body b is assumed.
	ICE	ICE = CH{b}, underlying body is assumed.
	ICE2	Ic + Ec = I + E 2C.

List of Figures

Figure ID: The first digit represents the chapter number; the rest of the Id represents the section and subsection number; the last letter, if any, differentiates the figures, in case there is more than one figure in the same chapter/section/subsection.

Figure Id	Description
1.5.24F	**Concurrent, independent internal ticks**: A body **can have** two concurrent, independent internal ticks.
1.5.24G	**Concurrent, independent external ticks**: A body **cannot have** two concurrent, independent external ticks.
1.5.24H	**Inherited ticks** are not independent and therefore cannot make a composite tick.
1.5.24J	**Atoms sharing an electron** only intermittently.
1.6	**Non-lineal ticks** and xp-ticks.
1.7	**General diagram** demonstrating most types of ticks.
3.06A	**Concurrent internal ticks** have a tree structure.
3.06B	**Concurrent external ticks** have a lineal structure.
3.07	**Venn Diagram** for different types of c-hulchuls.
3.10	**Comparison of ph[b] and ph[b].**
3.15.4	**ECH-equivalent bodies:** $ch[[b1]] \supseteq ch[[b]]$, where $b1 \subset b$.
4.3.2	**Composite ticks**: Limited inheritance of composite ticks.
4.9	**Time dilation using absolute motion.**
4.9.1	**Relativistic time dilation.**

LIST OF HULCHUL RELATED CONCEPTS

In Conceptual Order

Chapter 1

Inner-outer body relationship
    Inner body
    Outer body
    Lineal bodies
Inheritance of motion
The four types of bodies
    The U body
    Elementary Bodies
    Phantom bodies
    Common bodies
Outer body as a frame of reference
The three types of motion
    External motion of a body
    Internal motion of a body
    Composite motion of a body
Extended external motion of a body
Absolute motion of a body
Relative motion of two non-lineal bodies
Change of the three types of motion of a body

Prime ticks (p-ticks), the building blocks of motion
Chained ticks
Ticks
    Empty/non-empty ticks
    Lineal/non-lineal ticks

Length of a tick
Motion in terms of ticks
External tick
Internal tick
Composite tick
Pure external tick
Pure internal tick
Decomposition of a tick (Property #1 of ticks)
Inheritance of ticks—axiom (Property #2 of ticks)
Source tick
Destination (resultant inherited) tick
Range of lineal bodies
Independent ticks
Strong transitivity of ticks (property #3 of ticks)
Extended prime ticks (xp-ticks) (Property #4 of ticks)
Single body based ticks (Property #5 of ticks)
    Internal tick
    External tick
    Composite tick
Notation t[b, U]
Equation of a single body tick (Property #6 of ticks)
Native tick
Difference between motion of light/photons and elementary particles
Inheritance properties of pure prime ticks (Property #7 of ticks)
Composite tick may not be inherited as a composite tick
Inherited prime ticks may be chained ticks
Maximal ticks
    Maximal external tick
    Maximal internal tick
    Maximal internal tree of xp-ticks
Instants of motion (maximal external tick)
Axiom of external ticks (Property #8 of ticks)
Maximal external ticks (Property #9)
Axiom of exclusion (Property #10 of ticks)

Chapter 2

Order of events
Concurrency of events (c-event)
Set of events (e-set)
Concurrency disjoint (c-disjoint) e-sets
Concurrency equivalent (c-equivalent) e-sets
Prime hulchul (p-hulchul)
    External p-hulchul of a body b: ph[b]
    Relative p-hulchul of a body b in B: ph[b, B]
    Internal p-hulchul of a body b: ph(b)
    Pure external p-hulchul of a body b: ph[[b]]
    Pure internal p-hulchul of a body b: ph((b))
    Composite external p-hulchul of a body b: ph[E(b)]
    Composite internal p-hulchul of a body b: ph[I(b)]
    Total p-hulchul of a body b: ph{b}
Concurrency Id (CID)
Concurrency set (c-set)
Set of concurrency attributes: CIDS(S)
Concurrency operator: CNCY(P)
Concurrency difference operator: CDIF(P, Q)
Concurrency intersection operator: CINT(P, Q)
Concurrency hulchul (c-hulchul)
    External c-hulchul of a body b: ch[b] = CNCY(ph[b])
    Relative c-hulchul of a body b in B: ch[b, B] CNCY(ph[b, B])
    Internal c-hulchul of a body b: ch(b) = CNCY(ph(b))
    Pure external c-hulchul of a body b: ch[[b]] = CNCY(ph[[b]]
    Pure internal c-hulchul of a body b: ch((b)) = CNCY(ph((b)))
    Composite c-hulchul of a body b:
        ch[(b)] = CNCY(ph[E(b)]) = CNCY(ph[I(b)])
    Total c-hulchul of a body b: ch{b} = CNCY(ph{b})
Alternative formulas for different c-hulchuls

Chapter 3

Extended prime hulchul (xp-hulchul)
    External xp-hulchul of body b: xph[b]
    Internal xp-hulchul of body b: xph(b)

Relative xp-hulchul of body b in B: xph[b, B]
Composite external xp-hulchul of body b: xph[E(b)]
Composite internal xp-hulchul of body b: xph[I(b)]
Pure external xp-hulchul of body b: xph[[b]]
Pure internal xp-hulchul of body b: xph((b))
Total exp-hulchul of body b: xph{b}
Concurrency hulchul (c-hulchul)

> **Note**: c-hulchuls corresponding to xp-hulchuls are the same as those to p-hulchuls.

ch[b] = CNCY(xph[b]) = CNCY(ph[b])
ch[b, B] = CNCY(xph[b, B]) = CNCY(ph[b, B])
ch(b) = CNCY(xph(b)) = CNCY(ph(b))
ch[b] = CNCY(xph[E(b)]) = CNCY(ph[E(b)])
    = CNCY(xph[I(b)]) = CNCY(ph[I(b)])
ch[[b]] = CNCY(xph[[b]]) = CNCY(ph[[b]])
ch((b)) = CNCY(xph((b))) = CNCY(ph((b)))
ch{b} = CNCY(xph{b}) = CNCY(ph{b})

Universal c-ticks
Universal c-hulchul
Instants of time (universal c-tick)
Universal time as universal c-ticks
Body-specific time as internal c-ticks of the body
Hulchul domain (h-domain)
    Continuous h-domain
    Discontinuous h-domain
Density of h-domain
Counts of c-hulchuls
    PH[b] = count(ph[b])
    PH(b) = count(ph(b))
    PH[E(b)] = count(ph[E(b)])
    PH[I(b)] = count(ph[E(b)])
    PH[[b]] = count(ph[[b]])
    PH((b)) = count(ph((b))
    PH{b} = count(ph{b})

    XPH[b] = count(xph[b])
    XPH(b) = count(xph(b))

XPH[E(b)] = count(xph[E(b)])
XPH[I(b)] = count(xph[E(b)])
XPH[[b]] = count(xph[[b]])
XPH((b)) = count(xph((b))
XPH{b} = count(xph{b})

Ec = CH[b] = count(ch[b])
Ic = CH(b) = count(ch(b))
C = CH[(b)] = count(ch[(b)]
E = CH[[b]] = count(ch[[b]])
I = CH((b)) = count(ch((b)))

Density of time
Axiom 1 of elementary bodies
Axiom 2 of elementary bodies
Proposition #1:
     ch{b} = ch{q} = ch{U} = ch(U) for any bodies b and q
Universal constant: ch(U) for the same h-domain

Time outage
Uniform c-hulchul h-domain-wise
Middle bodies
Middle c-hulchul
CH-equivalent bodies
ECH-equivalent bodies
ICH-equivalent bodies
CCH-equivalent bodies

Chapter 4

Estimation of composite c-hulchul
Equation 2 of hulchul in I, C and E is untenable
Composite Concurrency pairs
Estimation of concurrency pairs
     The probabilistic method
     The simulation method
Limits on the values of I, E, C, Ic and Ec

$$\text{Maximum value of } C \leq \frac{IcEc(Ic + Ec)}{(Ic)^2 + (Ec)^2 + IcEc} \qquad (4.6A)$$

$$\text{Minimum value of } I \geq \frac{(Ic)^3}{(Ic)^2 + (Ec)^2 + IcEc} \qquad (4.6B)$$

$$\text{Minimum value of } E \geq \frac{(Ec)^3}{(Ic)^2 + (Ec)^2 + IcEc} \qquad (4.6C)$$

Proposition #2
Proposition #3

Glossary

Glossary of Terms Related to Hulchul

- We will define here only the terms related to hulchul.
- We will use the following information, by way of examples, in some of the definitions of the terms.

  $b1$, b, B, and $B1$ are bodies.

  $b1 \subset b \subset B \subset B1$.

  $T1 = t[b1, b]$, $T2 = t[b, B]$, $T3 = [b, B]$, $T4 = t[B, B1]$.

  $T12 = T1 + T2 = t[b1, B] = t[b1, b] + t[b, B]$.

  $T123 = T1 + T2 + T3 = t[b1, B1] = t[b1, b] + t[b, B] + t[B, B1]$.

** **Table (Glossary)**

Alternative formulas for c-hulchul:

$ch((b)) = CDIF(CNCY(ph(b)), CNCY(ph[b])$

$ch[[b]] = CDIF(CNCY(ph[b]), CNCY(ph(b))$

$ch[(b)] = CINT(CNCY(ph[b]), CNCY(ph(b))$

$= CINT(CNCY(ph(b)), CNCY(ph[b])$

Axiom 1 of elementary body: A body is either an elementary body or it has some elementary body as an inner body during each universal c-tick.

Axiom 2 of elementary body: An elementary body ticks during each universal c-tick; it may be an inherited tick or a native tick.

Axiom of concurrent external ticks: 1) An external tick of a body can't be repeated (replicated) concurrently. 2) A body cannot have two or more external ticks concurrently and independently unless their reference bodies are lineal. Suppose t1 = [b1, b] and t2 = t[b1, B]. If t1 and t2 are concurrent then b and B must be lineal. (All concurrent external ticks of a body form a maximal external tick of the body.)

Axiom of exclusion: The same tick cannot be both an internal and an external tick of the same body. An internal tick and an external tick of the same body must be independent. (This is the same as: An external tick of a body b cannot be inherited as an internal tick of b and an internal tick of a body b cannot be inherited as an external tick of b.)

Axiom of inheritance of internal ticks: A body b inherits all internal ticks of all inner bodies of b.

Axiom of inheritance of external ticks: A body b inherits all external ticks of all outer bodies of b.

Bodies, lineal: Bodies embedded one into another serially. b1, b, B and B1 are lineal bodies.

Body specific time (bs-time): See time.

Body, common: Any body other than an elementary body, a phantom body or the U body.

Body, elementary: A body, other than a phantom body, that has no inner body.

Body, four types: The U body, elementary body, phantom body, and common body.

Body, inner: See inner body.

Body, middle: *See* middle body.

Body, outer: *See* outer body

Body, phantom: A body p that can inherit neither external ticks of any outer body of p, nor internal ticks of any inner body of p, nor other bodies can inherit ticks of a phantom body.

Body, the U body: The universe as a whole.

BS-time. The same as body-specific time. See time.

C: Shorter notation for CH[(b)], the number of composite c-ticks of an underlying body b.

CCH-Equivalent bodies: Two bodies having the same value of composite c-hulchul but different values of internal c-hulchul or different values of external c-hulchul.

CDIF: Concurrency Difference operator: to find concurrency difference of two sets of events (e-sets). If P and Q are two e-sets, then CDIF(P, Q) = {x | x ∈ P and x is not concurrent with any y of Q}. It is similar to the difference of two sets but not the same.

C-Disjoint e-sets: The same as **concurrency disjoint.** Two sets of event (e-sets) will be called **C-Disjoint** If no member from one e-set is concurrent with any member of the other e-set. Example: In general, ph(b) and ph[b] are not c-disjoint, ch(b), ch[b] are not c-disjoint; ph((b)), ph[[b]] and ph[(b)] are c-disjoint, ch((b)), ch[[b]] and ch[(b)] are c-disjoint.

C-Equivalence of two e-sets: The same as **concurrency equivalent.** Two e-sets P and Q are **c-equivalent** if a member of either e-set is concurrent with some member of the other e-set. This means CNCY(P) = CNCY(Q). Examples: ph(b) and xph(b) are c-equivalent. In fact, any p-hulchul and its corresponding xp-hulchul of a body are c-equivalent.

C-Event: The same as **concurrency event**. It has a unique ID to be called concurrency ID (or CID). Example: All c-ticks are c-events.

Chained tick, external: A chained tick of the object body. T12 is an external chained tick of b1 in t[b1, B].

Chained tick, internal: As chained tick of the reference body. T12 is an internal chained tick of B in t[b1, B].

Chained tick: An internal or external chained tick. T12 and T123 are **chained ticks**. A tick that can be decomposed into two or more p-ticks (prime ticks). T12 and T123 are composite ticks. Length of a chained tick > 1.

CH-Equivalent bodies: Two bodies having the same values of corresponding c-hulchul for each type of c-hulchul.

C-Hulchul identity: $C(I+C+E) = (Ic)(Ec) - IE$.

C-Hulchul, composite: ch[(b)]: The set of common members of ch(b) and ch[b]. ch[(b)] = ch(b) ∩ ch[b]. ch[(b)] = CNCY(ph[(b)]) = CNCY(xph[(b)])

C-Hulchul, dissection of: See **Dissection.**

C-Hulchul, external: ch[b]: The set of classes of concurrent external xp-ticks (or p-ticks) of a body b. ch[b] = CNCY(ph[b]) = CNCY(xph[b]).

C-Hulchul, internal: ch(b): The set of classes of concurrent internal xp-ticks (or p-ticks) of a body b. ch(b) = CNCY(ph(b)) = CNCY(xph(b)).

C-Hulchul, middle: *See* middle c-hulchul.

C-Hulchul, pure external: ch[[b]]: The set of external c-ticks of a body b that are not internal c-ticks of the body b. ch[[b]] = ch[b] − ch(b) = ch[b] − ch[(b)]; ch[b] = ch[[b]] ∪ ch[(b)].

C-Hulchul, pure internal: ch((b)): The set of internal c-ticks of a body b that are not external c-ticks of the body b. ch((b))= ch(b) − ch[b] = ch(b) − ch[(b)]; ch(b) = ch((b)) ∪ ch[(b)].

C-hulchul, relative: ch[b, B] = CNCY(ph[b, B]) = CNCY(xph[b, B]), where b ⊂ B.

C-Hulchul, total: ch{b}: The union of internal c-hulchul and external c-hulchul of a body b. ch{b} = ch(b) ∪ ch[b] = ch((b)) ∪ ch[(b)] ∪ ch[[b]].

C-Hulchul, uniform: *See* uniform c-hulchul.

C-Hulchul, values of: Notation for counts of the six different types of c-hulchuls: I, E, C, Ic, Ec, ICE and ICE2.

C-Hulchul: The same as a **concurrency hulchul**. Any c-hulchul.

C-Pair: The same as a concurrency pair. A consecutive pair of concurrent internal and external c-ticks.

CID: Concurrency ID, a unique ID for each c-tick—kind of a unique time-stamp. All c-ticks belong to the same class, a class of CID's.

C-Instant: The same as **a concurrency instant:** The same as an external c-tick. An external c-tick of a body appears to be an instant of motion and an internal c-tick of a body appears to be an instant of time.

CINT: Concurrency Intersection operator: To find the concurrency intersection of two e-sets. If P and Q are two sets of events (e-sets), then CINT(P, Q) = {x | x ∈ P and x is concurrent with some y of Q}. It is similar to the intersection of two sets. In general, CINT(P, Q) and CINT(Q, P) are different. However,
CNCY(CINT(P, Q)) = CNCY(CINT(Q, P)).

CNCY: Concurrency operator: to find the set of concurrency events (c-set) from a set of events (e-set). Example: ch[b] = CNCY(ph[b]) = CNCY(xph[b]).

Composite tick: If a body b has both an internal tick and external tick concurrently, then the body b is said to have a **composite tick**. It is denoted by t[(b)]. Example: in the case of tick T123, each b and B has a composite tick.

Concurrency disjoint: The same as c-disjoint.

Concurrency Equivalence of two e-sets: See c-equivalence of two e-sets.

Concurrency hulchul: Same as c-hulchul. Any of several types of c-hulchuls.

Concurrency instant: See c-instant. Same as external concurrency tick (c-tick).

Concurrency Pair: The same as c-pair. A consecutive pair of concurrent internal and external c-ticks.

Concurrency Set: Same as c-set.

Counts of hulchuls: The count of a hulchul is denoted by replacing ph by PH, xph by XPH and ch by CH. Alternately, the count of a hulchul is denoted using the function count. For example, Count of ph[b] is denoted by PH[b] or count(ph[b]). As a special case: ch[b], ch(b), ch[[b]], ch[(b)], ch((b)) and ch{b} are also denoted by Ec, Ic, E, C, I and ICE respectively.

C-Set: A set of events such that no two of its events are concurrent. Same as concurrency set.

C-Tick, composite: A member of composite c-hulchul ch[(b)]. A composite c-tick is both an internal c-tick and an external c-tick of a body b.

C-Tick: The same as concurrency tick. A member of any c-hulchul.

Disjoint e-sets: Two e-sets are said to be disjoint if they are disjoint as sets; that is, they have no event in common. ph(b) and ph[b] are disjoint e-sets. xph(b) and xph[b] are disjoint e-sets. Two c-disjoint e-sets are also disjoint e-sets. Two disjoint e-sets may or may not be c-disjoint. Example: E-sets {e11, e12} and {f11, f13} are disjoint but they are not c-disjoint since both have the same c-set {<1>, <2>}. E-sets {e11, e12} and {f13, f14} are both disjoint and c-disjoint since their c-sets are {<1>, <2>} and {<3>, <4>} respectively.

Dissection of c-hulchul: A detailed analysis of all parts of c-hulchul and their relationships.

E: A shorter notation for CH[[b]], the count of pure external c-ticks in ch[[b]] for an underlying body b.

Ec: A shorter notation for CH[b], the count of external c-ticks in ch[b] for an underlying body b.

ECH-Equivalent bodies: Two bodies having the same value of external c-hulchul but different values of internal c-hulchul.

Embedded bodies: See lineal bodies.

Embedded ticks: See lineal ticks.

Empty tick: If b1 does not tick in b, then t[b1, b] is said to be an empty tick.

Equation 1 of c-hulchul: $I + C + E = ch(U)$ in terms of their counts and $I + C + E = 100\%$ in terms of their percentage for any body. $ch\{b\} = ch\{U\} = ch(b)$ for any body b.

Equation 2 of c-hulchul: This is a hypothetical equation thought to be true in the beginning but found untenable later after we found two bodies having the same value of external c-hulchul. The hypothetical equation is: $(I+C)^2 + (E+C)^2 = (ICE)^2$. Two independent equations in I, C and E are not feasible.

E-Set: A set of events.

Event: A tick is an example of an event.

Events, concurrency of: Notation for the order of occurrence of two events: $E1 \leftarrow E2$ denotes "E1 occurs earlier than E2". $E1 \rightarrow E2$ denotes "E1 occurs later than E2". $E1 \leftrightarrow E2$ denotes "E1 is concurrent with E2". The relations $\leftarrow$ and $\rightarrow$ are transitive. The relation $\leftrightarrow$ is reflexive, symmetric and transitive.

Events, order of: Notation for the order of occurrence of two events: $E1 \leftarrow E2$ denotes "E1 occurs earlier than E2". $E1 \rightarrow E2$ denotes "E1 occurs later than E2". $E1 \leftrightarrow E2$ denotes "E1 is concurrent with E2". The relations $\leftarrow$ and $\rightarrow$ are transitive. The relation $\leftrightarrow$ is reflexive, symmetric and transitive.

Extended prime hulchul: The same as xp-hulchul.
Any xp-hulchul.

Extended prime tick (xp-tick): The same as xp-tick. A prime tick of a body b or a prime tick of an inner or outer body of b. An xp-tick has property #4 of ticks.

Frame of reference: For the motion of a body b, we use an outer body of b as a frame of reference.

External motion: *See* motion.

H-domain, continuous: A continuous subsequence of universal c-hulchul ch(U). To be used like a continuous time interval.

H-domain, density of: Density of an h-domain h1 is the percentage of the number of c-ticks to the number of c-ticks in h1 as a continuous h-domain. The density of a continuous h-domain is 100%.

H-domain, discontinuous: A discontinuous subsequence of universal hulchul ch(U). To be used like a discontinuous time interval.

H-domain: A continuous or discontinuous subsequence of universal c-hulchul ch(U). To be used like a continuous or discontinuous time interval.

Hulchul domain: See h-domain.

Hulchul: Any prime hulchul (p-hulchul), any extended prime hulchul (xp-hulchul), or any concurrency hulchul (c-hulchul).

I: A shorter notation for CH((b)), the number of pure internal c-ticks of b, when the body b is known.

Ic: A shorter notation for CH(b), the number of internal c-ticks of b, when the body b is known.

ICE: A shorter notation for CH{b}, the number of all c-ticks of b, when b is known. ICE = I + C + E.

ICE2: ICE2 = Ic + Ec = I + 2C + E.

ICH-Equivalent bodies: Two bodies having the same value of internal c-hulchul but different values of external c-hulchul.

Idempotent operator: If the output after more than one application of an operator is the same as the output after one application of the operator, then CNCY is an idempotent operator because CNCY(CNCY(S)) = CNCY(S).

Independent ticks: Two ticks will be called **independent ticks** if neither is inherited from the other.

Inheritance of motion: Property of motion of a common or elementary body b that each inner body b1 of b goes wherever b goes whenever b1 ⊂ b. As a result, an inner body inherits the motion of all outer bodies and an outer body inherits the motion of all inner bodies.

Inheritance table: The table lists all possible types of the destination tick inherited from a given type of the source tick.

Inherited ticks, property of: If tick t1 is inherited from t2, then: t1 and t2 are non-empty, concurrent, logically equal, have the same number of prime ticks, and have different object bodies and/or different reference bodies.

Inner body: If body b1 is a constituent part of body b, then we will say b1 is an **inner body of b** and b is an **outer body of b1**.

Inner-outer body relationship: If body b1 is a constituent part of body b, then we will say b1 is an **inner body of b** and b is an **outer body of b1**. We will denote this relationship by b1 ⊂ b.

Instant of motion: We will treat maximal external ticks of a body b as **instants of motion** of b. There is one-to-one correspondence between maximal external ticks and external c-ticks of a body. All concurrent, independent external ticks of a body b form a maximal external tick of b, and therefore an instant of motion of b.

Internal motion: *See* motion.

Kinetic pulls: A maximal external tick of a body b, with length n > 0, exerts n kinetic pull(s) on b concurrently.

Length of a tick: The Number of concurrent independent prime ticks in a tick. It can be 0, 1 or > 1 and the tick is called an empty tick, or a prime tick or a composite accordingly.

Light, speed of: Light has the maximum speed because all ticks of light are native ticks in hulchul theory.

Lineal bodies: Two bodies are lineal if one is inner body of the other. Bodies embedded one into another serially. b1, b, B and B1 are lineal bodies.

Lineal ticks: One tick embedded into another tick. T1, T2 and T3 are **lineal ticks**. If two ticks are not lineal, then they are called **non-lineal ticks**. Two lineal ticks are parts of the same chained tick.

Maximal external tick: An external tick of a body with the maximum number of concurrent independent prime ticks is said to be a **maximal internal tick**. If t1 is a maximal external tick of b there exists B such that b ⊂ B and B has no external tick concurrently t1.

Maximal internal tick: An internal tick of a body with the maximum number of concurrent independent prime ticks is said to be a **maximal internal tick**. If t1 is a maximal internal tick of b there exists b1 such that b1 ⊂ B and b1 has no internal tick concurrently t1.

Maximal tree of internal p-ticks: All independent internal xp-ticks of a body b form a tree with b as the root, nodes as inner bodies of b, branches as internal xp-ticks of b, and leaves as inner bodies of b having no internal ticks.

Middle body: A body m with the property CH(m) = CH[m] is said to be a **middle body**. Middle bodies are possible since internal c-hulchul increases outer-body-ward and external c-hulchul increases inner-body-ward.

Middle c-hulchul: The c-hulchul of a middle body.

Motion in terms of ticks: Motion is a sequence of c-ticks. This is true for any type of motion.

Motion is like a movie: Motion is like a **movie**—a sequence of individual picture frames. Individual picture frames in this case are the instants of this kind of motion. Motion is not necessarily continuous; it is discontinuous in the case of common bodies.

Motion of photons: Photons have only native ticks, wavy straight motion, and the maximum speed.

Motion outage: Logically, during pure internal c-hulchul of a body b, b has **motion outage**. (We do not know if it has any physical significance except the obvious one. It is defined on the lines of time outage.)

Motion, absolute: The absolute motion of body b is the extended motion of b inclusive of all outer bodies of b.

Motion, change of type of: Change of motion of a body from one type to another type; types are: internal, external, composite, pure internal, pure external motion.

Motion, composite: Concurrent internal and external motion of a body.

Motion, difference between photon and elementary particle: In case of a photon, it is a sequence of native ticks. In case of an elementary particle, it is a sequence of native and/or inherited prime ticks.

Motion, extended: As an example, the combined motion of a moving train and the Earth is an extended motion of the train.

Motion, external: The motion of a body external to the body. If the body b1 moves in the body b, then it is an **external motion of b1** and internal motion of b. External motion of a body may be chained.

Motion, inheritance of: *See* inheritance of motion.

Motion, instant of: *See* instant of motion.

Motion, internal: Motion of a body internal to the body. If the body b1 moves in body b, then it is external motion of b1 and an **internal motion of b.** Internal motion of a body is the collective motion of all inner bodies of b with reference to b.

Motion, relative: If two bodies P and Q are moving relative to each other, how does one determine which body is at rest and which body is moving? Choose a common outer body R of P and Q. (The U body is always one of the choices in this case.) If P moves relative to R, then P moves. Similarly, if Q moves relative to R, then Q moves.

Motion, wavy straight: Elementary bodies and photons have **wavy straight motion** rather than straight motion because they move fast and consequently change direction fast. It is an implication of how we defined prime ticks.

Movie: Motion is like a movie—a sequence of individual picture frames. Individual picture frames in this case are the instants of this kind of motion.

Native tick: If an external prime tick t of a body b is not inherited, then it is called a **native tick.**

Non-empty tick: A tick t is **non-empty** if t is not empty.

Non-lineal bodies: When bodies are not lineal.

Non-lineal ticks: When ticks are not lineal.

Outer body: If b1 $\subset$ b, then b1 is **inner body of b** and b is an **outer body of b1**.

Particle accelerator: Proposition #2 appears to be the reason for the why a particle accelerator works.

Photons: Photons have: only native ticks, wavy straight motion, and the maximum speed. Photons neither inherit the motion of other bodies nor do other bodies inherit the motion of photons.

P-Hulchul, composite external: ph[E(b)]: The set of all independent external p-ticks of a body b, each concurrent with some internal p-tick of b. ph[E(b)] ⊂ ph[b]. ph[E(b)] and ph[I(b)] are disjoint since ph[b] and ph(b) are disjoint.

P-Hulchul, composite internal: ph[I(b)]: The set of all independent internal p-ticks of a body b, each concurrent with some external p-tick of b. ph[I(b)] ⊂ ph(b). ph[E(b)] and ph[I(b)] are disjoint since ph[b] and ph(b) are disjoint.

P-Hulchul, external: ph[b]: The set of all independent external p-ticks of a body b.

P-Hulchul, internal: ph(b): The set of all independent internal p-ticks of a body b.

P-Hulchul, pure external: ph[[b]]: The set of all independent external p-ticks of a body b not concurrent with any internal p-tick of b. ph[[b]] ⊂ ph[b].

P-Hulchul, pure internal: ph((b)): The set of all independent internal p-ticks of a body b not concurrent with any external p-ticks of b. ph((b)) ⊂ ph(b).

P-hulchul, relative: ph[b, B], where b ⊂ B: The set of all independent prime ticks of b in B.

P-Hulchul, total: ph{b}: The union of internal p-hulchul and external p-hulchul of a body b is called the **total p-hulchul** of b. (**Note:** ph(b) and ph[b] are disjoint but not c-disjoint.) ph{b} = ph(b) ∪ ph[b].

P-Hulchul: Any p-hulchul. (The *same as* prime hulchul.)

Prime hulchul: The same as p-hulchul. Any prime hulchul—external, internal, composite, pure external, pure internal, relative or total.

Prime tick: The same as p-tick. The smallest possible movement of any body will be called a **prime tick (or p-tick)**. After each p-tick, a body pauses and/or changes direction. "Pause and/or change direction" appears to be the reason for wavy straight, rather than straight, motion of fast moving bodies such as photon or a material elementary particle.

A prime tick is to motion what a photon is to light or what an elementary particle is to matter.

Principle of relativity: All laws of physics are the same in all uniformly moving frames of reference. This principle for motion can be derived from the inheritance of motion.

Probabilistic method: The probabilistic method used to estimate the number of composite c-ticks, ch[(b)], of a body b. The output of this method is close to the output of the Simulation Method.

Properties of ticks: *See* ticks.

Proposition #1: Total c-hulchuls of all bodies are the same for a given h-domain.

Proposition #2: External c-hulchul, internal c-hulchul, pure external c-hulchul and pure internal c-hulchul of a common body are less than 100% and greater than 0% for all sufficiently large h-domains. (Note: ch(b) + ch[[b]] = ch[b] + ch((b)) = 100%.). This proposition appears to be the reason why a particle accelerator works. It may also be responsible for the presence of shells/sub-shells in an atom.

Proposition #3: Light and photons (phantom bodies) can neither inherit internal or external motion of any body nor can any body inherit motion of phantom bodies.

P-Tick, composite: A member of composite p-hulchul ph[(b)]. A pair of concurrent internal and external p-ticks of b.

P-Tick, external: As a p-tick of the object body of a p-tick or a member of external p-hulchul ph[b]. If t = t[b, B], b ⊂ B, then t is an external p-tick of b1 and an internal p-tick of b.

P-Tick, internal: As p-tick of the reference body of a p-tick or a member of internal p-hulchul ph(b). If t = t[b, B] is a prime tick, b ⊂ B, then t is an external p-tick of b1 and an internal p-tick of b.

P-Tick, pure external: As a pure external p-tick of the object body of a tick. Or a member of pure external p-hulchul ph[[b]]. An external p-tick of a body b not concurrent with any internal p-tick of b.

P-Tick, pure internal: As a pure p-tick of the reference body of a tick or a member of pure internal p-hulchul ph((b)). An internal p-tick of a body b not concurrent with any external p-tick of b.

P-Tick: The same as prime tick. The smallest possible movement of any body will be called a **prime tick (or p-tick)**. After each p-tick, a body pauses and/or changes direction. "Pause and/or change direction" appears to be the reason for wavy straight, rather than straight, motion of a photon or a material elementary particle.

Pure Ticks: A pure internal tick or pure external tick. If a body has an external tick but no current internal tick, then the external tick will be called a **pure external tick**. If a body has an internal tick but no concurrent external tick, then the internal tick will be called a **pure internal tick**.

Range of lineal bodies—1st half: The range of lineal bodies between an elementary body and a middle body. In this range, for any body b, count(ch(b)) ≤ count(ch[b]).

Range of lineal bodies—2nd half: The range of lineal bodies between a middle body and the U body. In this range, for any body b, count(ch(b)) ≥ count(ch[b]).

Simulation method: The simulation method used to estimate number of composite c-ticks of a body. The output of this method is close to the output of the Probabilistic Method.

State of motion: State of motion can be relative or absolute.

State of rest: State of rest can only be relative. No body is at rest absolutely except possibly the U body. Each body has internal or external motion at all instances of time.

Tick, destination: If tick t2 is inherited from tick t1, then t1 is called a source tick and t2 is called a destination tick.

Tick, empty: If b1 does not tick in b, then t[b1, b] is said to be an **empty tick**.

Tick, external: As the tick of the object body of a tick. T1 is an external tick of b1. If t1 = t[b1, b] then t1 is an external tick of b1 and an internal trick of b.

Tick, internal: As the tick of the reference body of a tick. T1 is an internal tick of b. If t1 = t[b1, b] then t1 is an external tick of b1 and an internal trick of b.

Tick, non-empty: A tick t is **non-empty** if t is not empty.

Tick, relative: t[b1, b] is a **relative tick** of b1 in b. In this case, b1 ticks in b.

Tick, source: If tick t2 is inherited from tick t1, then t1 is called a source tick and t2 is called a destination tick.

Tick: Any prime tick (p-tick), composite tick, extended prime tick (xp-rick), or any concurrency tick (c-tick), or an empty tick. **A tick is an event**.

Ticks, independent: *See* independent ticks.

Ticks, inherited: *See* inherited ticks.

Ticks, property #1 (Decomposition): A composite tick can be decomposed into all p-ticks.

Ticks, property #2 (inheritance): Axioms of Inheritance of ticks.

Ticks, property #3 (Strong transitivity of ticks): Suppose b1 ⊂ b ⊂ B. If t[b1, b] **or** t[b, B], then t[b1, B] and vice versa.

Ticks, property #4 (Extended prime ticks—xp ticks): A p-tick of an outer body of b is called an **external xp-tick** of b. A p-tick of an inner body of b is called an **internal xp-tick** of b. An xp-tick of a body b is either a p-tick of b or concurrent with a p-tick of b.

Ticks, property #5 (Single body based ticks): If t[b1, b] is a non-empty tick, then b1 has an external prime tick, represented as **t[b1]**, and b has an internal prime tick, represented as **t(b)**. **t[(b)]** represents a composite tick of b. If t[b1, b] is a chained tick, then there exists at least one b0 such that b1 ⊂ b0 ⊂ b and b0 has a composite tick, represented as **t[(b)]**; that is, b0 has an internal tick and an external tick concurrently.

Ticks, property #6 (Equation of a single tick): If b1 ⊂ b, then t[b1] = t[b1, b] ∪ t[b]. If b1 ⊂ b and b1 ticks, then either b1 ticks in b and/or b ticks.

Ticks, property #7: Inheritance of pure ticks as pure ticks of the same type.

Ticks, property #8: See Axiom of concurrent external ticks

Ticks, property #9: Concurrent external ticks of a body form a maximal external tick of b. (A maximal external tick of a body b is also called an instant of motion of b.)

Ticks, property #10 (Axiom of exclusion). See axiom of exclusion.

Time dilation: Physical phenomenon when time of a body slows down as the speed of a body increases. Time dilation appears to be caused by time outage.

Time outage: A body does not experience time during pure external c-hulchul of a common body, a phenomenon to be called **time outage**. This appears to be the cause of time dilation. Pure external c-hulchul is intermittent and consequently time outage is intermittent—intermittency may be to any degree of fineness.

Time, body-specific or bs-time: Internal c-hulchul, ch(b), of a body will be called **body-specific time or bs-time**. As a special case, bs-time for the U body will be called **universal time or u-time**.

Time, density of: Density of time is the density of bs-time of a body b. Density of bs-time of a body is the density of ch(b) as an h-domain. Density of u-time is 100%. Only the U body can have 100% density of time.

Time, elapsed: In computer jargon, this is clock time. Processing time is CPU time. CPU time is similar to the body-specific time and universal time is similar to elapsed time.

Time, instant of: Universal c-ticks are instants of time in terms of hulchul.

Time, processing: In computer jargon, this is CPU time. Elapsed time is clock time. CPU time is similar to body-specific time and universal time is similar to elapsed time.

Time, universal: The same as u-time. ch(U) is the universal time. Body-specific time of a body b is ch(b).

Uniform c-hulchul: A body b has uniform c-hulchul h-domain-wise if the ratio of c-hulchul value of each of I, C and E of b to the total c-hulchul, ICE, of b is the same for all large h-domains. For example, the value of I/ICE remains the same h-domain-wise. However, values of I/ICE, E/ICE and C/ICE may be all different.

Universal concurrency hulchul: ch(U): *See* c-hulchul.

Universal constant: The count of total c-hulchul for the same h-domain is a universal constant since its value is independent of a body.

Universal c-tick: A member of ch(U).

Universal time: *See* time.

Wavy straight motion: *See* motion.

XP-Hulchul, external: xph[b]: The set of all independent external xp-ticks of a body.

XP-Hulchul, internal: xph(b) = The set of all independent internal xp-ticks of a body.

XP-hulchul, relative: xph[b, B] where $b \subset B$: xph[b, B] = The set of all independent xp-ticks of b in B.

XP-Hulchul, pure external: xph[[b]] = {x | x $\in$ xph[b] and x is not concurrent with any member of xph(b)}.

XP-Hulchul, pure internal: xph((b)) = {x | x $\in$ xph(b) and x is not concurrent with any member of xph[b]}.

XP-Hulchul, composite external: xph[E(b)] = {x | x ∈ xph[b] and x is concurrent with some member of xph(b).

XP-Hulchul, composite internal: xph[I(b)] = {x | x ∈ xph(b) and x is concurrent with some member of xph[b]}.

XP-Hulchul, total: xph{b} = Union of internal xp-hulchul and external xp-hulchul of a body b is called the **total xp-hulchul** of b. (**Note**: xph(b) and xph[b] are disjoint but not c-disjoint.) xph{b} = xph(b) ∪ xph[b].

XP-Hulchul: Any xp-hulchul. The same as an extended prime hulchul.

XP-Tick, external: An external prime tick of a body b or an external prime tick of an outer body of b.

XP-Tick, internal: An internal prime tick of a body b or an internal prime tick of an inner body of b.

XP-Tick: The same as an extended prime tick. A member of any XP-hulchul. A prime tick of a body b or a prime tick of an inner or outer body of b. An xp-tick has the property #4 of ticks.

BIBLIOGRAPHY

Asimov, Isaac. 1992. Atom: Journey Across The Subatomic Cosmos. Penguin Books USA Inc., New York, NY.

Cole, George. 2007. Wandering Stars. Imperial College Press. 57 Shelton Street, Covent Garden, London WC2H 9HE.

Einstein, Albert. 1990. The Meaning of Relativity. Princeton University Press, Princeton, New Jersey.

Einstein, Albert. 1010. Relativity: The Special & the General Relativity. 2010. Martino Publishing, Mansfield, CT.

Feynman, Richard. 1988. QED: The Strange Theory of Light and Matter. Princeton University Press, Princeton, New Jersey.

Frank, Adam. 2011. About Time. Free Press, New York, NY.

Hart-Davis, Adam. 2011. The Book of Time. Firefly Books(U.S.) Inc. Richmond Hill, Buffalo, New York.

Hawkins, Stephen. 1998. A Brief History of Time. Bantam Books, New York, NY.

Newton, Isaac. 1999. The Principia: Mathematical Principles of Natural Philosophy. University of California Press, Berkeley and Los Angeles, CA.

Peebles, P. J. E. 1993. Principles of Physical Cosmology. Princeton University Press, Princeton, New Jersey.

Roy, A. E., 2005. Orbital Motion, Taylor and Francis Group, New York, NY

Vries, Hans de. 2009, Understanding Relativistic Quantum Field Theory.

INDEXES

- **Frequent references to a term**: In such a case, generally, only the page number of the page, where the term is defined, is listed.
- **Infrequent references to a term**: In such a case, generally, all distinct page numbers for all references are listed.